MATTER AND ENERGY

MAXWELL PRESS

MATTER AND ENERGY

Frederick Soddy M.A., F.R.S.
Winner of Nobel Prize for Physics
Lecturer in Physical Chemistry and Radioactivity
University of Glasgow

MAXWELL PRESS

Chennai Trichy NewDelhi

MAXWELL PRESS

ISBN 978-93-90063-47-5 Maxwell Press

No. 44, Nallathambi Street,
Printed and bound in India Triplicane, Chennai 600 005

MJP 893

Publisher : C. Janarthanan

Publisher's Note

The legacy of a country is in its varied cultural heritage, historical literature, developments in the field of economy and science. The top nations in the world are competing in the field of science, economy and literature. This vast legacy has to be conserved and documented so that it can be bestowed to the future generation. The knowledge of this legacy is slowly getting perished in the present generation due to lack of documentation.

Keeping this in mind, the concern with retrospective acquiring of rare books has been accented recently by the burgeoning reprint industry. Maxwell Press is gratified to retrieve the rare collections with a view to bring back those books that were landmarks in their time.

In this effort, a series of rare books would be republished under the banner, "Maxwell Press". The books in the reprint series have been carefully selected for their contemporary usefulness as well as their historical importance within the intellectual. We reconstruct the book with slight enhancements made for better presentation, without affecting the contents of the original edition.

Most of the works selected for republishing covers a huge range of subjects, from history to anthropology. We believe this reprint edition will be a service to the numerous researchers and practitioners active in this fascinating field. We allow readers to experience the wonder of peering into a scholarly work of the highest order and seminal significance.

MAXWELL PRESS

CONTENTS

PERIODIC TABLE

Group O.	Group I.	Group II.	Group III.	Group IV.	Group V.
Helium He 3·99	Lithium Li 6·94	Beryllium Be 9·1	Boron B 11·0	Carbon C 12·00	Nitrogen N 14·01
Neon Ne 20·2	Sodium Na 23·00	Magnesium Mg 24·32	Aluminium Al 27·1	Silicon Si 28·3	Phosphorus P 31·04
Argon A 39·88	Potassium K 39·10	Calcium Ca 40·07	Scandium Sc 44·1	Titanium Ti 48·1	{ Vanadium V 51·0
	Copper Cu 63·57	Zinc Zn 65·37	Gallium Ga 69·9	Germanium } Ge 72·5	Arsenic As 74·96
Krypton Kr 82·92	Rubidium Rb 58·45	Strontium Sr 87·63	Yttrium Yt 89·0	Zirconium Zr 90·6	{ Niobium Nb 93·5
	Silver Ag 107·88	Cadmium Cd 112·40	Indium In 114·8	Tin } Sn 119·0	Antimony Sb 120·2
Xenon Xe 130·2	Caesium Cs 132·81	Barium Ba 137·37	[Lanthanum La 139·0		Cerium Ce 140·25
	Europium Eu 152·0		Gadolinium Gd 157·3		Terbium Tb 159·2
	Thulium Tm 168·5		Ytterbium Yb 172·0	Lutecium] Lu 174·0	{ Tantalum Ta 181·5
	Gold Au 197·2	Mercury Hg 200·6	Thallium Tl 204·0	Lead } Pb 207·10	Bismuth Bi 208·0
Radium Emanation 222·		Radium Ra 226·4		Thorium Th 232·4	

OF THE ELEMENTS

Group VI.	Group VII.	Group VIII.		
Oxygen O 16·00	Fluorine F 19·0			
Sulphur S 32·07	Chlorine Cl 35·46			
Chromium Cr 52·0	Manganese Mn 54·93	Iron Fe 55·84	Cobalt Co 58·97	Nickel Ni 58·68
Selenium Se 79·2	Bromine Br 79·92			
Molybdenum Mo 96·0	—	Ruthenium Ru 101·7	Rhodium Rh 102·9	Palladium Pd 106·7
Tellurium Te 127·5	Iodine I 126·92			
Praesodymium Pr 140·6		Neodymium Nd 144·3	Samarium Sa 150·4	
Dysprosium Dy 162·5		Erbium Er 167·7		
Tungsten W 184·0	—	Osmium Os 190·9	Iridium Ir 193·1	Platinum Pt 195·2
(Polonium)				
Uranium U 238·5				

MATTER AND ENERGY

CHAPTER I

PHYSICAL HISTORY

THE behaviour of matter and energy represents one aspect only of human knowledge, which is generally known by the name of physical science. It seems well to state at the outset that, throughout these pages, when the term science is employed it refers solely to this one branch. Physical science enjoys the distinction of being the most fundamental of the experimental sciences, and its laws are obeyed universally, so far as is known, not merely by inanimate things, but also by living organisms, in their minutest parts, as single individuals, and also as whole communities. It results from this that, however complicated a series of phenomena may be and however many other sciences may enter into its complete presentation, the purely physical aspect, or the application of the known laws of matter and energy, can always be legitimately separated from the other aspects. This aspect comes first, not necessarily in relative importance, but in the order of the scientific definition of the phenomena and of the problems it presents for a solution. A great simplification thereby results, which is too often neglected. Complete ignorance of these laws is, nowadays, rare, for they enter into the general common sense of the age, and any flagrant violation of them is quickly exposed. But the neglect to give precedence to the purely physical aspect of the complicated occurrences and events of human experience in their orderly presentation, has led to much confused history and a general lack of clearness as to the precise terms with Nature on which the race exists on this planet. There is a special branch of study known as physical geography, but the need for a similar branch of physical history does not appear to have been widely felt. The laws expressing the relations between energy and matter are, however, not solely of importance in pure science. They necessarily come first in order, in the fundamental sense described, in the whole record of human experience, and they control, in the last resort, the rise or fall of political systems, the freedom or bondage of nations, the movements of commerce and industry, the origin of wealth and poverty, and the general physical welfare of the race. If this has been too imperfectly recognized in the past, there is no excuse, now that these physical laws have become incorporated into everyday habits of thought, for neglecting to consider them first in questions relating to the future. It is an interesting and by no means hackneyed side of the subject to consider, so far as the operation of purely physical laws can teach, exactly what the future has in store for this world and the complicated civilisation that it contains. Is it a stable and permanent movement, or does it carry in itself, like the life of the individuals that comprise it, the seeds of its own inevitable decay? Moreover, if, as will transpire when the nature of the controlling physical laws has been made clear, it is ephemeral and will decline the sooner the more rapid its development and the more glorious the zenith it attains, what alteration of the existing conditions would suffice to convert it into a physically stable and permanent movement? On these great questions, rendered the more fascinating because of the disposition, since the development of the doctrine of evolution, to consider the fate and future of the individual as of little importance compared with the fate and future of the species, physical science in its later developments has much to say that is of general interest. The proverb counselling the cobbler to stick to his last is a good one; but since the province of physical science is the universe and all that moves therein, its right to be heard first, in order of presentation of the subject only, cannot be withstood. It may or may not assist in disclosing the fundamental bearings of any question, but anything it has to say will in general be definite and, in so far as the laws are perfectly known, incapable of being invalidated by any other considerations whatever. The laws may not be fully known and may give rise to false deductions, a case of which arose in the question of the duration of geological time. In such a case, the discussion of the conflicting evidence can only result in the advance of knowledge. Physical science, by reason of the universality of its laws, has something to say on almost every subject. It need only be stated once for all, that although the purely physical side can be considered separately, it does not render other points of view less necessary, though, of course, it is only with the physical point of view that the present volume is concerned. To adopt for the moment the language of Spencer's *Classification of the Sciences*, referred to in the Introductory volume of this Series (p. 89), physical science supplies subject-matter for every actual occurrence in the universe, but none of the truths outside of physical science can help in the solution of physical problems.

The recognition of the fundamental physical conditions which control the destinies of a race, too often occurs too late in its development to be of service. History throws some strange sidelights on this blindness to the obvious. The upward progress of the race has, for example, been classified into succeeding eras, each designated by the name of a material. Thus are distinguished the Stone Age, the Bronze Age, the Iron Age, and the Steel Age. The names indicate that the era in question was associated with a certain degree of mastery over a particular material sufficient to enable new weapons to be forged in the struggle for existence. Yet, when the early records of these eras are examined, little or nothing is found about the pioneers whose knowledge and craft effected these broad advances. Often were they held in such contempt that it was considered almost beneath the dignity of an educated man even to make himself superficially acquainted with the technical processes to which, in the judgment of history, his era owed its initiation. To come to more recent times, how many people blessed with a liberal education would be at a loss if asked offhand what steel is, and how it is distinguished from iron; or would recognise even the names of the great founders of the modern era?

Fundamental as materials are in shaping the broad lines of progress, it is necessary to go but very little deeper to come upon something equally fundamental but less obviously so. Materials are employed merely as weapons, tools, or instruments for the utilisation of power or energy. Even the food we eat is not the end but the means of living. Life is physically distinguished from death by movement, and what food is to the motion of living organisms, fuel is to the motion of mechanical engines. With the advent of steel the utilisation of the natural sources of energy has progressed with enormous strides. Less than a hundred years ago little was known about energy, and, indeed, the modern idea of energy as a definite fundamental existence was not developed till well on in the last century. Isolated examples of energy, apart from that of living beings, have been known necessarily and some have been utilised from the remotest times. The wind that propels the sailing ship was probably one of the first forms to be harnessed to the affairs of life. The phenomena of fire, and the thermal energy derived from it, were known to all but the most primitive races, though its recognition as one of the manifestations of energy is not yet a century old. Not until the law of the conservation of energy was established, and it was shown that energy like matter is indestructible and uncreatable, could energy be regarded as one of the fundamental physical existences. Its recognition, as a separate entity, distinguishes the present age from all its predecessors. This is the Age of Energy, or rather this is the beginning of the ages of energy, the Age of the Energy of Coal. The triumphs of this age have been sung in season and out of season. Already, however, science has outgrown such immature jubilation. That this still is the age of the energy of coal is unfortunately only too true, and the whole earth is rendered the filthier thereby. Moreover, the age will last just so long as the coal supply lasts, and after that the last state of the race will be worse than the first, unless it has learned better. Only ten years ago the prospect was, in fact, anything but a cause for jubilation; but these last years have wrought a wonderful revolution in our knowledge of energy, and therefore in the future outlook of the race, now entirely bound up with that of energy. It is possible to look forward to a time, which may await the world, when this grimy age of fuel will seem as truly a beginning of the mastery of energy as the rude stone age of paleolithic man now appears as the beginning of the mastery of matter. It may await the world, but by no means of necessity awaits it. The prospect is physically possible, but the realisation depends also upon man and whether he can ever hope to rise to the heights of knowledge the problem demands. The discoveries in connection with the recently explored field of radioactivity have put an entirely different complexion on the question as to how long the energy resources of the world may be expected to last. It has transpired that there exists in matter, associated with its ultimate atoms, that is, by definition, with the smallest particles capable of separate existence of the elements or most fundamental known forms of matter, sufficient potential energy to supply the uttermost ambitions of the race for cosmical epochs of time. But, just in proportion as the prizes to be won by the progressive mastery over the physical universe become the more magnificent, the more does their achievement transcend in difficulty and seeming impossibility the older successes. Think of the ages that elapsed after man kindled his first fire, before the world hummed to the tune of the steam engine. Think of the ages that preceded even this remote discovery, so ancient that no record of it survives, during which in natural conflagrations man must have been made aware of the energy in fuel before he had learned how to liberate it at will. With reference to the newly revealed stores of atomic energy in matter, and to the time, so lightly pictured in the imagination, when it may raise the race to the loftiest pinnacle of its ambitions, we are in the position that savage man bears to the present, aware, but in every other way ignorant. True we are thoroughly familiar with the more superficial processes of nature, and bring to the task a trained and disciplined intelligence. But the task increases in proportion to the knowledge which defines it.

Practically King Coal is as likely as ever to die naturally of exhaustion as to be deposed by another monarch; and, if so, he carries away with him the means of subsistence of our boasted civilisation. But the recognition of the boundless and inexhaustible energy of Nature, and the intellectual pleasure and gratification it affords, brightens the whole outlook of the twentieth century.

Mere accumulations of knowledge, sifted, classified, and reduced to their final most concise expression in a series of text-books, are little more than the sepulchral monument of science. However complete and accurate, mere knowledge deadens rather than develops the intellect. The history of the winning of knowledge preserves a sparkle of life and may stimulate as well as instruct. But the real value of science is in the getting, and those who have tasted the pleasure of discovery alone know what science is. A problem solved is dead. A world without problems to be solved would be devoid of science though it might be full of scientific text-books and dictionaries. Such is the prospect that recent discoveries are opening up that there is no fear that science will yet awhile be sighing, like Alexander, for fresh worlds to conquer.

Before the doctrine of its conservation was established, energy was mysterious and unaccountable in its comings and goings. To-day it is no longer a mystery. The unaccounted-for appearance or disappearance of a quantity of energy in any process, however complex, would rouse as much scientific interest as the mysterious appearance or disappearance of matter. When it appears it must come from somewhere, and when it disappears it must go somewhere. Gradually this Law of Conservation has supplied the physicist with an experimental test of reality in a changing universe. What appears and disappears mysteriously, giving no clue of its origin or destination, is outside of his province. To him it has no physical existence. What is conserved has physical existence, whether it is tangible and ponderable like matter, or intangible and imponderable like energy. Early writers, when they really meant what is now called energy, often used the term force; and the idea of force, as will later be discussed, has confused the issues and retarded the growth of science to an almost incalculable extent. Carlyle says, meaning energy—"Force, force, everywhere Force; we ourselves a mysterious Force in the centre of that 'There is not a leaf rotting on the highway but has Force in it: how else could it rot?'" The very idea of Force is, however, what would be termed an anthropomorphism, that is to say, it ascribes the behaviour of inanimate objects to causes derived from the behaviour of human beings. We have come to associate the motion of matter with somebody or something pulling or pushing it. When one body is observed to move towards another, like a stone falling to the ground, it has been supposed that, although no agent is visible, something must be pulling it. What, however, is actually observed is a change of position of the body, which acquires at the same time motion or velocity. The observation is correctly expressed by saying that energy, before associated with the position of one body with reference to another (potential energy), has changed into energy of motion (kinetic energy). To suppose that the one body attracts or pulls the other with a certain "force" is to imagine a cause, which if it existed would account for the effect. Forces are not conserved, they have no physical existence, but they still survive even in scientific parlance, mainly because of the poverty of the language, which hardly allows effects to be expressed without some causal inference. They are bad gates through which to approach the study of energy, as is evident from the fact that mechanics existed in a highly developed state for centuries before the discovery of the conservation of energy. In mechanics, which is the science of the motion or absence of motion of matter in bulk, forces have a definite meaning, and in terms of energy they are measured by the change in the kinetic (or potential) energy of a body when its position changes by the unit of length. Many of the most important changes of energy are due to changes of position, too small to be measurable, between the smallest particles of the substance. The energy changes are, however, easily measurable. The attempt even to imagine forces to exist in such cases as the causes of the changes of energy, in absence of all knowledge, not only of the actual distances involved but also of the variation of the imagined cause with the distance, is to invent an elaborate, perfectly vague and befogging mode of expression for a very simple effect. It is better to try to grasp the meaning of energy as a fundamental fact of experience than to begin, with totally inadequate knowledge, to derive from the action of living beings a shallow analogy which, if true, would serve as a possible explanation of a few of the grosser manifestations in which energy plays a part.

Energy is recognised in two forms, kinetic and potential. The first depends on motion, the second on the position of the body under consideration, and the law of conservation states that any loss of energy of motion is balanced by a gain due to position, and *vice versa.* But it is possible to select cases in which the distribution of the kinetic energy changes, as among various moving bodies, without any abiding changes of potential energy. The effect of position can thus be eliminated, and the question reduced to its simplest form. The law of conservation, then, has reference simply to kinetic energy or energy of motion. The question that first has to be asked is, What is conserved? Neither motion as such, nor what Newton termed quantity of motion, or momentum, the product of the mass of the moving body and its velocity, is conserved. A sufficiently good example is in the collision of two elastic balls. No material is perfectly elastic, it is true, and in all actual collisions some of the energy of motion of the body as a whole is transformed into heat, or the energy of motion of its smallest parts with reference to one another. This part can be accurately measured in practice, so that for the purposes of a simple illustration it is legitimate to consider the balls chosen as perfectly elastic, colliding on a level plane—for example, a billiard table. If two perfectly elastic balls collide, no matter what the relative masses of the balls, or what their relative velocities, there is only one quantity, involving these masses and velocities, which is the same before and after the collision. That quantity is the sum obtained by adding together the product of the mass of each ball and the square of its velocity. The measure of kinetic energy adopted is half the mass multiplied by the square of the velocity. The numerical factor one-half is not of great significance in the present connection. The important fact is that it is the square of the velocity and not the velocity itself which is conserved. It is the same in the case of all phenomena in which pure motion uninfluenced in any lasting way by position is considered. No matter what changes have occurred in the relative motions of the moving parts, the sum of the products of the masses into the square of their velocities is conserved. This therefore answers the question as to how kinetic energy depends on motion. It might be supposed that a truth deduced thus haltingly from the behaviour of imaginary perfectly elastic substances had no very great application in the real world. As a matter of fact, it is what we call imperfect elasticity which has no great range of application in the real world, the world of molecules, as contrasted with the gross world of matter in bulk, which is all our unaided senses can perceive. The application of the principle to all cases of pure motion is universal, and the reservation as to imperfect elasticity had to be made simply because it is not immediately obvious that loss of kinetic energy of motion and its transformation into heat is merely subdivision of motion among smaller particles of matter that can be directly perceived. Molecules, if they are really the smallest particles of matter that exist, must be perfectly elastic, as later on will be quite evident. It would be as absurd to postulate an inelastic molecule in pure science, as it is at present understood, as it would be to assume for the motion of each individual in a surging crowd the general chaotic aimlessness which appears to characterise the whole. The science of heat is mainly one grand general example of the very case that has been postulated, namely, that of pure motion uninfluenced by position, not in the visible seeming world of gross masses, but in the invisible real world of molecules.

In physics, work and energy are interchangeable terms. The simplest case of doing work is the lifting of a weight from the ground to a height. The amount of work done, and the amount of energy spent in doing it, are simply proportional, first, to the mass or quantity of the matter lifted; second, to the height it is lifted. Mass is practically measured by weight, so long as the measurements refer to one part of the earth. Owing to the fact that the position of an object on the surface of the earth relative to its centre varies with the latitude, the distance apart being appreciably greater at the equator than at the poles, a given mass weighs slightly less, and falls to the ground a little less rapidly, in the tropics than elsewhere. What follows refers to a single locality where weight is truly a measure of mass. It does not require any appreciably different amount of work to lift a weight in the upper room of a house than in the basement. When a weight is lifted, kinetic energy disappears and the equivalent quantity of potential energy, measured by the weight of matter multiplied by the height it is lifted, is produced. A foot-pound is one unit of potential energy due to height. Twenty foot-pounds is practically the same whether it refers to the work done on a 20 lb. weight raised one foot, or on a 1 lb. weight raised 20 feet. When the weight falls again, the potential energy disappears and the equivalent quantity of kinetic energy, measured by half the mass of matter multiplied by the square of the velocity it acquires at the end of its descent, reappears. The truth of the statement, derived from experimental observation, that it is the square of the velocity, not the velocity itself, which is a measure of the kinetic energy, may now be made a little more obvious. The kinetic energy acquired by a falling weight is the equivalent of the potential energy it possessed prior to falling, and is therefore the product of the weight and the height of fall. If kinetic energy were proportional to the velocity simply, just as potential energy is proportional to the height, it would follow, therefore, that the velocity of a falling body should increase uniformly with the height it falls. Whereas the velocity increases, as every one knows, uniformly with the time taken for the fall. As the speed gets faster and faster a greater distance is traversed in each succeeding second, and therefore for each succeeding foot fallen through the velocity must increase by a less and less amount. The velocity acquired is proportional not to the height, but to the square root of the height of fall. The kinetic energy acquired is proportional to the height, and must therefore increase according to the square of the velocity. If further illustration were needed that the kinetic energy acquired by a falling body is proportional simply to the height of the fall, all that is necessary is to carry out the fall in two equal stages. The body falls the same height in each stage, and therefore acquires the same velocity and kinetic energy. But in falling the whole way the velocity acquired is not twice that acquired in falling half-way, but the square root of twice. The law of dependence of kinetic energy on motion can thus be deduced from the observed laws of falling bodies.

Space does not permit more than a reference to the early history of the doctrine of energy. The first conception had reference to chemical energy, and was contained in the Theory of Phlogiston which dominated chemistry during the eighteenth century. The twin laws of conservation, that of matter and of energy, struggled competitively for birth in the search of science after the unchanging entities. Energy was always being confounded with matter. Even the great chemist Boyle, who gave us the modern conception of elements, thought that heat was ponderable. The search was intuitively after conservation, and matter was less elusive than energy. Hence everything intuitively believed to be real ran the risk of being regarded as material. This was the fate of phlogiston. As first put forward, phlogiston was something which escaped during fire, or combustion, with the light and heat evolved. It was a pure anticipation of what is now called energy. Combustible substances were regarded as rich in phlogiston. The lode-star of conservation appeared first in the following way. When various prodnets of processes, which were recognised as being analogous to combustion, like the calces, or as we should say oxides, of the metals, or sulphuric acid and the sulphates, were heated with highly inflammable bodies like coal, oil, or organic matter, the original combustible substances, that is, the metals or the sulphur, as the case might be, were regenerated. This view recognised that during combustion something (energy), which manifested itself as light and heat, escaped;

and that, before the original materials could be got back from the products of combustion, this something had to be put back. It was only recognised much later that during combustion something *material* (oxygen) was absorbed from the air, and that before the products could be regenerated the compound formed had to be decomposed and the oxygen liberated. The spirit of chemistry tended towards pure materialism. The later followers of the phlogiston theory made the fatal mistake of materialising phlogiston. With the enthronement of the balance, and the test of weight as the criterion of material reality, the existence of phlogiston as a material substance was disproved, and the theory itself fell into quite undeserved disrepute. In its original form it anticipated by more than a century the modern doctrine of energy. It is most wonderful to reflect that the first idea of conservation in science arose not in connection with weighable matter, but with the elusive, imponderable energy. The second or modern phase arose in connection with the nature of heat, after the law of the conservation of matter had been established, when it was no longer possible to regard heat as a material fluid. Davy and Rumford, at the commencement of last century, both had the modern conception of heat as a mode of motion of matter, and both came very near to establishing it. The latter was engaged in the boring of cannon by means of horses, and observed the large amount of heat continuously generated during the operation. He records in one experiment that by the work of a single horse 19 lbs. of water were boiled, and the cannon, drill, and all the machinery employed were heated up to the boiling point of water in 140 minutes. But boring changes the state of the metal from the compact to the finely divided form of borings and turnings, and it had to be proved that the latter had not a less capacity for heat than the former. Joule, who repeated the experiments in another form by merely churning water, which suffers thereby no physical change except rise of temperature, established the modern view that the source of the heat is in the power or energy expended. He measured exactly how much heat is produced by a given amount of mechanical work, and found that 772 foot-pounds have to be spent to raise the temperature of 1 lb. of water 1° Fahrenheit. The water of a waterfall, 772 feet high, tumbling over into a deep pool, so that practically all of its kinetic energy is converted into heat, is 1° F. hotter at the bottom than at the top. Expressed in the modern scientific units, which are based on the gram (.035 oz. or 15.4 grains) as the unit of mass, the centimetre (0.394 inch) as the unit of length, and the second as the unit of time, Joule's equivalent is 42,650. That is, a weight of 42.65 kilograms falling a centimetre has sufficient energy to raise the temperature of 1 gram of water 1° Centigrade. The latter unit of heat is known as the caloric. In the combustion of coal, considered as pure carbon, the heat evolved would raise the temperature of a mass of water about 8000 times that of the coal 1° Centigrade, or a mass 14,000 times 1° Fahrenheit.

Conversely, when work is produced by any heat engine the equivalent quantity of heat disappears. As is well known, the conversion of heat into mechanical work is a very wasteful process. But if it were possible to convert the chemical energy of coal completely into work, without first burning it to liberate the energy as heat, the energy of 1 ton of coal would then be sufficient to lift one of the largest liners, weighing 20,000 tons, 500 feet high. In other words, the chemical energy of coal is equivalent to that of a mass equal to the mass of the coal falling under gravity a distance of 2000 miles, or one quarter of the earth's diameter. The engineer's unit—the horse-power—defined after actual tests with the best British cart-horses, is a rate at which work is done or energy is spent. One horse-power will lift a weight of 550 lb. one foot per second. A horse-power-hour, or one horse-power acting for a period of one hour, is thus almost exactly two million foot-pounds. The "Board of Trade Unit," at which electrical power is supplied to consumers, is the "Kilowatt-hour," and is about a third greater than the horse-power-hour. This is the energy consumed by an ordinary 4-lamp electric radiator in an hour, but if converted into mechanical energy instead of heat it would raise a weight of 1 ton, 1200 feet high.

The Apostle Paul had no thought of physical things in his mind when he used the words, "The things which are seen are temporal, but the things which are not seen are eternal." But the words can be applied with profit to illustrate, perhaps more forcibly than any single sentence, the essential nature of energy. It is only the temporary changes in the form and relative amount of energy which are manifest. So long as energy neither changes in amount nor position in space, it belongs to the unseen and eternal. No direct evidence of its existence can be obtained. It is true that, if in similar cases changes in the energy have occurred, the existence of the energy may be deduced from analogy. Thus so long as nitroglycerin or guncotton do not suffer changes there is no possible way of measuring or recognising the energy they contain, though from the observation that, in previous cases, such substances have under certain conditions invariably been found suddenly to evolve a large amount of energy, suffering at the same time complete change of material character, the existence of the energy may be inferred with certainty. The total energy contained in matter depends upon the extent to which it can be changed. When it has been pulled to pieces and its smallest component parts dispersed out of the range of one another's influence, the difference between the energy initially and finally present can be found. The absolute value of potential energy involves a complete knowledge of the real, as contrasted with the so far attained, limits of subdivision of materials. The absolute value of kinetic energy involves a knowledge of the real or absolute velocity of the moving body, as contrasted with the motion relative to the earth, which is all that usually has to be taken into account.

But the fact that the energy resources of Nature may be practically limitless does not in the least affect the way the accounts are kept, or the amount that can be derived for practical purposes. This amount ultimately controls the number of people the world can support, and the intensity of their struggle for existence. The discovery of a large quantity of gold leaves the world not a penny the richer. Those that find it are, by convention, made wealthy at the public expense. In addition, if gold increases in abundance without a corresponding increase in prosperity, the purchasing power of money diminishes, prices of all commodities rise, and long and bitter industrial strife results between the wage-earner and the employer. But energy and wealth are synonymous. Energy is the thing of which gold itself is but the guinea stamp, to adapt the simile of Burns. A find of energy in Nature means an addition to the general wealth, a postponement of the day of bankruptcy, which each new invention of science, on the other hand, brings nearer. Physical history, the science of the physical conditions underlying the past, present, and future history of the race, has some rude disillusionments in store. It was not so very long ago that a Member of Parliament condoned the wasteful methods of utilising coal on the ground of the additional benefit that accrued thereby to the coal industry. At the moment of writing, employers and employed in that industry are calmly contemplating a complete stoppage while they fight out their differences, and the public experiences a temporary return to the original conditions of life of primitive man. It is curious that it should have taken so long for the age to realise that it differs physically and fundamentally from all the preceding eras of history only in the utilisation of the energy of fuel. The ignorance and apathy of the ancients in reference to the metals, the mastery over which distinguished them from their predecessors, is paralleled by the situation to-day with regard to this new life-blood of civilisation. The horror of the waste of food is inborn, the horror of the waste of fuel has still to be acquired.

Let us spend a few moments in examining critically the conditions of the existence of the present age of energy. What is this coal, which is rated so lightly as a mere commodity, to be bought and sold, wasted or utilised, as much or as little as individuals think fit? As commonly regarded, it is a substance which, by indomitable industry, perseverance, and pluck, man succeeds in digging up from the bowels of the earth. The enterprise and expansion of the world, consequent upon an abundant coal supply, are regarded as the just return of human effort and expenditure. Its market price, at the pit's head, of so many shillings per ton, represents the cost of bringing it to the surface, just as its price in a locality remote from the mine represents this cost together with the cost of transportation. Not a thought is given as to what the coal has cost Nature. Science regards coal in a different light, as a legacy from the bygone past, which man, having attained his majority and acquired a knowledge of its utilisation, is engaged in spending and dissipating as fast as it can be brought to the surface. The cost at the pit's head no more has any relation to the real value of the coal than the cost of its transport above the ground. The only part of its value for which any return is demanded is the cost of transportation from the place where it occurs to the place where it is used. The coal itself represents the accumulation of solar energy over almost incredible periods of time. The primæval forests of the carboniferous era were not, the geologists tell us, mushroom growths. The layers of shale between each shallow seam indicate that, alternately, the site of the forest was dry land and the bed of the sea. For the formation even of a single seam, alterations of the contour of land and sea must have occurred which require not years nor lifetimes, but geological epochs. It is not possible to reckon what the cost of coal has been in the economy of Nature, nor how many ages of future time will be necessary to recuperate the amount now burned in a single year. On this irreplaceable and diminishing commodity the future prospects of civilisation depend. Primitive races were dependent entirely on the regular day to day supply of natural energy, and were imperilled by each temporary failure or diversion of the supply. All that civilisation has yet contrived is to augment the natural supplies out of capital, under the delusion that it was supporting itself out of income on a more liberal scale than its predecessors, because of the greater scope of its knowledge and intellectual powers! It is a pure delusion. The knowledge which shall provide as well as spend, and which shall place civilisation upon the broad flowing river of energy, which supplies the vast requirements of the universe over infinite periods of time, is not yet born. The age in which we live, the age of coal, draws its vivifying stream from a dwindling puddle left between the comings and goings of the cosmical tide.

CHAPTER II

MATTER: I. ATOMS AND MOLECULES

We possess five senses by means of which the mind makes its acquaintance with the external world. Possibly the mind has other channels of communication not comprised within the five senses; but if so, they have not yet come within the scope of science. In the history of the science of matter two main types of mind can, in general, be distinguished, one of which has developed into the modern type of scientific mind, although the other type is by no means non-existent. The two types may be illustrated by the ancient and modern conception of elements. The word element is possibly derived from the letters *l m n* of the alphabet, conveying the idea that material things were built up of elements as a word is built up of letters. But the ancient and modern ideas of these fundamental things called elements were entirely different. The older type of mind regarded all the forms of matter as a combination of certain elementary qualities, the newer type as a combination of certain elementary substances.

There seems to have been distinguished by the Greek mind two fundamental qualities and their opposites, the one Dryness, the other Hotness, with their opposites Wetness and Coldness. These four qualities were combined in pairs to make the four "Elements," Earth, Air, Fire, and Water. "Fire" was Hotness combined with Dryness; "Earth," Coldness and Dryness; "Water," Coldness and Wetness; "Air," Hotness and Wetness. These so-called elements had no relation to the actual things called earth, air, fire, and water beyond the fact that the ancients considered the actual things typical examples of all the possible combinations of the qualities they thought most fundamental. Probably many have puzzled themselves over the meaning of the "four Elements," Earth, Air, Fire, and Water. They are scarcely worth it. The type of mind which created the four Elements was in the ascendant all through the Dark Ages, and as a direct consequence arose Alchemy—the beginnings of Chemistry. If the qualities of things are regarded as more fundamental than the things themselves; or if things are looked upon as having no existence apart from the qualities which they possess, the transmutation of one element into another appears very much the same as any other kind of chemical change. The ancients were acquainted with brass, and with the manner of its production by heating copper with ores of Zinc. Here we have a case, they might argue, where one of the qualities of copper, namely, its colour, has been changed into that of gold. This is the first step in the transmutation of copper into gold. Why should it not be possible, one at a time, to change each property of copper, in the same way as its colour, so as ultimately to arrive at a substance having all the qualities of gold? On their philosophy brass was gold so far as its colour went, not, as we should say, like it.

The modern habit of thought recognises things as having a real existence apart altogether from the particular qualities or properties by means of which the things make themselves known to the five senses. The acceptance of this habit of thought among scientific men has been due mainly, not to any formal proof, but to its fertility and to the undoubted value of the results which follow from it. Deep down somewhere in the processes of thought the ultimate test of reality appears to be the Law of Conservation. Does the soul exist? If so, it must be immortal. Is matter real or a mere impression of the mind? It cannot be created or destroyed, and therefore has an existence apart from the mind. Lastly, has energy a specific existence, or is it merely a convenient abstraction? Energy is conserved like matter, and therefore obeys this test of objective existence. Now consider the essential quality of all the phenomena of the occult world—apparitions, ghosts, spirits, astral bodies, and so on. It is that they are not conserved. They appear no one knows whence, and they disappear no one knows whither. They do not obey this test of reality, and the doubt remains how far they exist apart from the brain which apprehends them. It would ill become science to deny the reality of things with which it has no concern. But it has most definitely, but at the same time unconsciously, limited itself till now to the recognition of only those existences which appear to obey the Law of Conservation, and for which, therefore, there is this much of proof of reality.

What, then, are the fundamental existences recognised by physical science? Probably there are at least three:—Matter, Electricity, and Energy. The commonest questions ever asked are—"What is Matter?" "What is Electricity?" "What is Energy?" The questions reveal the intuitive method of the human mind in approaching Nature—the attempt to explain things not understood in terms of things simpler and more fundamental. Obviously, however, it cannot be applied to the most fundamental conceptions themselves. To be asked to define or explain a fundamental entity is to be asked to explain bricks in terms of houses. It is just because science has asked these questions over and over again without getting an answer that the existence of these three things is considered fundamental. A legitimate question would be, "How is a thunderstorm explained in terms of electricity, matter, and energy?" But to ask this question, "What is Matter?" is to assume that there exists something more simple and fundamental than matter, out of which it is in some way made up. There may be, but if so it is not yet completely known. We make up our minds, therefore, to accept the existence of as few fundamental things as possible which cannot be explained in terms of anything else. When it becomes possible to say exactly what they are, they will no longer be fundamental, but there always must be a certain number of fundamental things not explainable in terms of anything else, representing the limits to which the analysis of phenomena has been reduced.

Now the human mind intuitively believes that there are very few fundamental things in this sense, and if experience shows that there are more, it is apt to anticipate future discovery by assuming the complexity of what has to be regarded strictly as fundamental. This is precisely the case with matter. So far matter has been referred to as one of the fundamental existences, and it has been tactily assumed that all matter is made up of various combinations of the same unknown "primary stuff" or "protyle." As a matter of fact, before the discovery of radioactivity, no less than eighty elements were known, none of which could be formed from or turned into any other, and each was therefore a fundamental form of matter, having a separate existence. Just as at present it would be futile to ask what is matter, so it would be to ask what is gold, or copper, or any other of the eighty elements. Each is a separate and distinct thing, and has to be accepted as such. All its properties may be learnt, studied, and classified, but not one can be explained.

Every known material thing in the universe may be regarded as composed of one or more of these elements in definite quantities, capable of being analysed or decomposed into its several constituent elements, and often of being re-formed from them. The complicated materials making up the bodies of living beings are easy to decompose into their constituent elements, but only in certain cases have they been built up artificially or "synthesised" directly from the elements. By far the greater number of the important products of the vegetable world have now been synthesised, one of the more recent being camphor, which engaged the attention of numerous chemists for many years before it was artificially prepared. The synthetical production of rubber has long been known. Often, however, it is immensely more costly to make the products artificially than to grow them, so that it must not be supposed that the synthesis of any commodity means the immediate death of the natural industry. In the case of the madder plant, the natural source of the dye alizarin, however, this has been the result. A visitor to the South of France, surprised at the absence of madder plantations where once they had flourished, was told, when he asked the reason, that "they made it now by machinery." In the mineral world no materials are known of which the composition has not been ascertained, and a great number of the special crystallised forms of minerals, including the gems, ruby and diamond, have been artificially produced, though in the latter case, so far, only as microscopic crystals. The progress of synthesis, or the building up of natural materials from their constituent elements, proceeds apace. Even some of the simpler albuminoids, a class of substances of great importance in the life process, have recently been artificially prepared. None of the known materials are regarded in this respect as peculiar. The idea that a peculiar "vital force" acted in the chemistry of life is extinct. Some of the natural products are known to be exceedingly complicated in their structure, and their synthesis is therefore proportionally difficult. But progress is steady. The modern chemist builds up the compounds occurring in nature from their constituent elements as an architect causes a house to be built up out of bricks and mortar. The elements are the bricks, and energy may be regarded as somewhat akin to the mortar. Naturally occurring substances form, however, only one class. Innumerable entirely new compounds have been produced in the last century. The artificial dye-stuffs, prepared from materials occurring in coal-tar, make the natural colours blush. Saccharin, which is hundreds of times sweeter than sugar, is a purely artificial substance. New explosives, drugs, alloys, photographic substances, essences, scents, solvents, and detergents are being poured out in a continuous stream. The philosophic blemish that there are still eighty fundamental forms of matter instead of one, is compensated for on the practical and utilitarian side. The possible new combinations of these eighty elements among themselves are still legion in spite of the innumerable compounds already prepared. What wonder if chemistry amid this wealth of materials had become unphilosophical and had almost ceased to ask what are these elements! The very idea of transmutation was gradually coming to be looked upon as a lingering survival of an ancient heresy, when, quite unexpectedly, it was demonstrated that the radioactive elements were transmuting themselves. However, these developments need not be considered till later. The best introduction to the science of matter is got by making up the mind frankly to accept the separate existence of almost fourscore different elements, each something of a law unto itself, having, it is true, most marked and important analogies with the others, but not directly derivable from or transformable into any of them.

Now, if we follow another ancient philosophical line of thought, we shall arrive at the first most important property distinguishing these elements, namely, the relative weights of their atoms. The ancients asked themselves the question whether matter occupied space continuously or discontinuously. If you took a piece of solid matter, or a drop of liquid, and divided it into two equal parts and then subdivided one of these parts in the same way again and again, can you imagine the process going on for ever, or would you at last arrive at the "atom," as they called it, the particle so small that it could not be further divided? Science has definitely adopted the view of the "discrete" or "grained" structure of matter as opposed to the view that it occupies space continuously. The properties of gases make it the only possible view to adopt. Gases differ from liquids and solids in expanding into and uniformly filling every empty space that offers itself, and from the laws they obey it is possible not only to prove definitely that they consist of small particles moving about freely in space with great velocity, but even to count the number of particles contained in any quantity of a gas. These particles are not called "atoms," however, but "molecules." Let us suppose that we had chosen for the subdivision process, imagined by the Greeks, a drop of liquid water—not, that is to say, one of the fundamental elementary substances, but a compound made up of the two elements, hydrogen and oxygen. On the adopted view, at some stage in its subdivision we should come on the single unit particle, or molecule, of water, the smallest particle of water that can exist. It could not properly be described as the atom of water, because although it represented the limit of conceivable mechanical subdivision, it can still be decomposed, like any other quantity of water, into its constituent elements, hydrogen and oxygen, by chemical methods. The word molecule has thus been employed to represent the smallest single particle of any substance, whether elementary or compound, that has a separate existence; whilst the word "atom" has been used to denote a constituent elementary particle of a molecule. Thus the molecule of water is known to consist of two atoms of hydrogen and one of oxygen. The conclusion might be jumped at that, for an element, the molecule and the atom are the same. In rare cases they are, but more often they are not. The atoms possess certain powers of combining with other atoms, as the existence of compound substances shows. If they cannot get other atoms to unite with they will unite among themselves. Single atoms do not, as a rule, continue to exist free for any appreciable length of time. At the moment the molecule is broken up they do undoubtedly start existence single, but at the first opportunity they unite. This accounts for the fact that many of the elements at the moment of their formation, or in the "nascent state" as it is termed, often exhibit extraordinary reactivity and bring about many reactions which the same elements in the ordinary state are quite unable to effect. The molecules of hydrogen, oxygen, and nitrogen, and most of the elementary gases, consist of two atoms of these elements. But another gaseous form of oxygen, called ozone, is known, the molecule of which consists of three atoms. It is very different from ordinary oxygen, having a pungent smell and being extremely reactive, so that it instantly oxidises, or burns up, many materials not attacked by oxygen at ordinary temperature. The third atom is in the proverbial position of not being wanted, and at the first opportunity takes another partner. The ozone molecule separates into an ordinary oxygen molecule and a single free, and therefore exceedingly active oxygen atom. Then, too, there is the case of the diamond. Diamond is the purest natural form of the element carbon, and it has been suggested that its hardness, great density, etc., are due to six atoms of carbon being combined in a ring to form the molecule,—a form of union which is a favourite one for this element in compounds. Ozone and diamond are both converted into the commoner varieties, oxygen and carbon, by the simple action of heat alone, which proves their identity with these elements.

The rare gases of the atmosphere—helium,

argon, and the rest of the family—are unique in possessing no power of combination. Hence they do not combine with themselves, and their molecules are composed of single atoms, or, as it is termed, are "monatomic." They are, speaking anthropomorphically, the misanthropes among matter. Mercury in the form of vapour is the only known similar case, under ordinary circumstances, though at a very high temperature many gaseous elementary substances are more or less completely dissociated into single atoms.

Properties which furnished the basis of the Aristotelian system of elements depend not only on the constituent materials, but on the manner in which their atoms are arranged in space among themselves. When compounds are considered, there often exist several entirely different compounds, having the same kind and same number of atoms in their molecules, differently arranged. These are termed "isomers."

There is, however, much more involved in the modern terms atom and molecule than could ever be derived from the purely deductive reasoning of ancient philosophers. Indeed these conceptions would have very little scientific value if it were not for one very important consideration included in the modern use of the terms. If any one kind of elementary or compound substance is considered, the atoms or molecules, respectively, are all precisely alike down to the minutest particular. So far as the atoms are concerned this holds with the utmost rigidity. As regards the molecules, there are certain substances, called "tautomeric," for which it is believed that the molecule vibrates between two separate configurations. For these, of course, the statement has to be suitably qualified. The most striking evidence of this uniformity, as regards atoms, is furnished by the sharpness of the lines of light constituting the spectrum of the element. These lines, as will be shown, must originate from the vibrations of electrically charged systems associated with the atoms; and if the vibrations of different atoms were in the least out of tune with one another, the lines would be blurred and diffuse. Professor Schuster has calculated, from the sharpness of the spectrum lines, that the separate periods of vibrations among the atoms of an element do not differ from one another by *so much* as would be represented, in a collection of clocks, by one out of every eight losing or gaining a second in every twenty-three days. But chemistry tells the same story. It draws its elementary materials from widely different localities and kinds of compounds. Yet the elements show no differences among themselves, such as might be explained by slight differences in the individual atoms. Again, the spectroscope tells us that, in the most distant stars, the same elements exist as here, and that the periods of the vibrations which cause them to emit light are identical with those of their terrestrial representatives.

These profound conclusions forced themselves first on chemists in their study of chemical composition. When carbon or charcoal is burnt in air or oxygen, there may be two different kinds of compounds produced. If there is plenty of oxygen, the well-known gas, popularly called carbonic acid gas, and, more strictly, carbon dioxide, is formed. But with a deficiency of air, the highly poisonous and entirely different gas, employed in lethal chambers, and known as carbon monoxide, results. The first contains 27.28% carbon by weight and 72.72% of oxygen, whilst the second contains 42.86% of carbon and 57.14% of oxygen. Quite a large number of commonly occurring compounds had been analysed by chemists and the composition expressed percentage-wise as above, before it occurred to Dalton, the founder of the modern Atomic Theory, to express the results not as percentages, but in terms of a unit of one of the constituents. Thus the above results might be expressed as the amount of oxygen combined with one gram, ounce, pound, or any convenient unit of weight of carbon. Then it is seen that, whilst carbon monoxide contains per gram of carbon about $1\frac{1}{4}$ grams of oxygen, carbon dioxide contains about $2\frac{1}{2}$ grams. In the first case the amount of oxygen is exactly twice that in the second. This "Law of Multiple Proportions" was found to extend to all compounds in which the same elements united in different proportions. There was only one conclusion possible. The composition by weight of compounds is fixed and definite. Therefore the atoms, or smallest particles which combine together, must themselves have fixed and definite weight. Moreover, these atoms combine as units. One atom of carbon, as in the present case, may combine with one or with two atoms of oxygen, but not with one atom and a fraction. Henceforth the conception of the atom became more than a mere philosophical deduction. It is a necessity to explain the experimental facts of chemical composition. It would take us too far out of our course to consider at any length in what way this conception has grown until we are as convinced of the existence of these atoms and of their uniformity and invariability as if we could count and measure them. Indeed, they are actually counted in certain special cases in radioactivity. But something must be said as to the manner in which this work has culminated in assigning to each element a number representing the relative weight of its atom in terms of that of some standard element.

The proportions by weight of the elements in compounds give us the relative weights of the atoms of the elements, provided always that we can determine the relative number of atoms of each element in the molecule of the compound. Thus, in the compound carbon monoxide, the oxygen weighs $1\frac{1}{4}$ times as much as the carbon. Before we can conclude that the single atom of oxygen weighs $1\frac{1}{4}$ times that of the single atom of carbon, we must know that in the compound, carbon monoxide, there are equal numbers of the two kinds of atoms. Dalton guessed this and happened to be correct, but many of his other guesses were less happy. Really, quite a number of steps are involved before relative atomic weights can be deduced from the chemical composition. The first step is to find the relative weights of the molecules of as many of the compounds of the element as possible. This apparently difficult task is in reality very easily accomplished, because of a fundamental generalisation first made in the beginning of the last century. Equal volumes of all gases at the same temperature and pressure contain the same number of molecules. In this form it has been known successively as Avogadro's rule, Avogadro's hypothesis, and, lastly, Avogadro's law. In the first instance the view was put forward to harmonise Dalton's theory of atoms with some very simple general relations, discovered by Gay-Lussac, between the volumes of gases which combine together and the volumes of the products resulting, when these also are gaseous. If, for example, the volumes of hydrogen and oxygen are measured which combine together, when the mixture is exploded by a spark or a flame, to form water, and all the measurements are made above 100° C., so that the water produced by the combination remains gaseous, as steam, it is found that the volume of hydrogen is almost exactly twice that of the oxygen, and is equal to that of the steam resulting. This law of simple combining volumes, like the law of Multiple Proportions, holds for all combinations between gases. It led Avogadro to make his famous suggestion. For, on his view, the example taken would simply mean that two molecules of hydrogen unite with one molecule of oxygen to form two molecules of water. If the molecule of hydrogen contains the same number of atoms as the molecule of oxygen, this means, therefore, that one atom of oxygen combines with two atoms of hydrogen. Since 9 parts by weight of water contain 1 part of hydrogen and 8 parts of oxygen by weight, the atomic weight of oxygen in terms of that of hydrogen, as 1, is approximately 16. Gradually all the assumptions have been eliminated, and it has been found that a variety of different methods confirm the value of the atomic weights obtained in this way. Mainly for the convenience of analysts the unit adopted is oxygen, the atomic weight of which is made exactly 16. On this scale that of hydrogen is 1.008.

CHAPTER III

MATTER: II. THE ELEMENTS

THE special science of matter, or chemistry, occupies at the present time a curious intermediate position. On the one hand, it is connected with the more philosophical science of physics, or, to use an older term, natural philosophy, which tirelessly seeks to represent or "explain" natural phenomena, such as light, heat, and electricity, with the minimum number of fundamental assumptions, and those the simplest and most probable it is possible to conceive. On the other hand, it is connected with the descriptive sciences, embraced under the old term natural history, which find ready to hand in Nature a wealth of forms and modifications, connected by more or less close analogies and relationships, divided into genera, species, and varieties, and in which the power of artificially reproducing the work of Nature is either very limited or does not exist. Natural philosophy may "explain" a rainbow, but not a rabbit. A rainbow can be constructed at will out of perfectly colourless beginnings, which is a proof that there is no secret about a rainbow as such. But nothing but rabbits will or can produce a rabbit, a proof again that we cannot say what a rabbit is, though we may have a perfect knowledge of every anatomical and microscopic detail. Applying the same argument to matter, we may have a perfect knowledge of every reaction and property of an element, may know the wave length of every line in its spectrum with almost incredible accuracy, may have studied every possible compound of it, but still we do not know what a single element really is. In this sense chemistry is a purely descriptive science. Again in the other direction, in the chemistry of the excessively unstable complicated compounds which constitute the living body, the science is again descriptive, although in this direction every year the philosophical domain is extended. There seems no limit to possibilities of its extension, and the time has gone by when it seemed impious even to suggest the bare possibility of our one day being able to synthesise food-stuffs apart from the life process. Between these two domains, and within these limits, the science of matter is as philosophical as physics. The inquiring mind is more or less satisfied with the general explanation of the countless materials which exist. We can form mental pictures of their inner construction and can put them together artificially, in many cases by processes more efficient and economical than the natural ones. Will this ever be true of the elements themselves? Between the completely satisfying explanation of things and the mind in ignorance so profound that it does not know there is anything possible to be known, is a wide region, and somewhere in this gap the science of the elements at present remains. There are two possible methods of advance. One is to loose the reins of a brilliant imagination, to let it go for a space untrammelled by the limitations of knowledge, and only, at the end, bring the consequences of the process to be confirmed or rejected by fact. This is a method which, traditionally, has been unduly discouraged by chemists, for it is rare that the most erroneous theory, if original, and capable of experimental test, does not result sooner or later in a substantial increase in experimental knowledge. The other method is, as in all the purely descriptive sciences, to evolve by some conprehensive scheme order and definiteness out of the confused fabric of resemblances and differences, similarities and contrarieties, among the units which make up the whole. These eighty or more elements, each of which is a fundamental separate existence, exhibit among themselves the most remarkable set of resemblances and differences, and may be divided into groups or families. The beginning of a comprehensive scheme expressing these relationships dates from the middle of last century, when an Englishman, J. A. R. Newlands, revealed a connection between the properties of the chemical elements and the weights of their atoms which is of a most curious and fascinating kind. He seems to have been a man of little force of character and inexperience, mistaking scepticism for criticism, and discouraged thereby. Scepticism is always with us, but real criticism is the rarest thing in science; for criticism involves knowledge more even than discovery. Newlands was a pioneer, and mistakenly in his own region he gave to authority the deference which only belongs to its own. In consequence, his idea was not properly developed until it occurred independently to two other celebrated chemists, Mendelejeff and Lothar Meyer.

If the elements are written down in order of their atomic weights, beginning with the smallest, the successive members will, as a rule, differ abruptly from one another in general character, like the notes in a scale of music; but after a certain number have been set down, corresponding with the number of notes in an octave, the general character of the elements already set down will be reproduced in the same order by another set. It happened that in Newlands' time the number of elements that had to be passed through before the initial character repeated itself was seven, which is the same as the number of notes in the octave, and hence he called the generalisation the Law of Octaves. Owing to the discovery of the new family of similar elements in the atmosphere, beginning with argon, another note, so to speak, has to be added to the scale, making eight. The Periodic Classification, in one of the various forms in which it is now usually represented, is shown in the Table (pp. 6–7). The elements are arranged in order of atomic weight in horizontal columns of eight, as a rule, the elements falling under each other in vertical columns being analogous to each other. Each vertical column is therefore a family group and is indicated by the number 0 to 7, according to its position. The group number corresponds, as a rule, with the valency of an element, or the number of units of combining power it possesses. The zero group are nonvalent, and have no combining power whatever. It comprises the rare gases found in the atmosphere whose molecules exist as single atoms without power of uniting together. Group I contains the monovalent metals of the alkalies; these unite with one other monovalent atom, such as chlorine. Thus common salt is a compound of one atom of sodium and one of chlorine. Group II contains the divalent metals of the alkaline-earth family, of which calcium is the commonest. Common lime is a compound of one atom of calcium and one of the divalent element oxygen. Group III contains the trivalent earth elements, of which alumina, the oxide of aluminium, a compound containing two atoms of aluminium and three of oxygen, is typical. The next group are tetravalent, and contain some of the most important elements, like carbon and silicon, the oxides of which have two atoms of oxygen to one of the element. After this the elements usually act with several valencies forming different classes of compounds, and the group number represents the maximum valency. The most common valency is 8 minus the group number, so that Group V is generally trivalent, Group VI divalent and Group VII monovalent again. Groups I to III comprise the metals which, combined with oxygen, form alkalies or bases. The last three groups contain the non-metals which, when united with oxygen, form the acidic oxides. Group V contains nitrogen and phosphorus, which with carbon, hydrogen, and oxygen play the largest part in the constitution of living matter. Group VI contains the sulphur elements in addition to oxygen, and Group VII the halogen elements, fluorine, chlorine, bromine, and iodine. In general it may be said that the greater the difference between the group numbers of two elements, the greater the probability that they will combine together energetically and form well-defined compounds not easy to decompose again. In the Table, the symbol of the element and the value of its atomic weight are placed below it. A line in a blank space indicates that probably a still undiscovered element exists there.

The Periodic Classification is, however, by no means so simple or perfect as has been indicated. The simple period, or "octave" of eight elements, applies strictly only to the first sixteen elements, which are in many ways rather different from the subsequent heavier elements. To understand the Periodic Classification fully would require an intimate detailed knowledge of the individual character of each of the known elements. The more highly trained a chemist is in this direction the more will he see to interest and attract his attention in the Periodic system. Its apparent defects as well as its obvious advantages will give food for thought probably for a long time to come. Strictly speaking, it has no defects; for it is no theory, but a simple mode of expressing the facts. Like the elements it represents, it is a law unto itself. It cannot be explained, but only described and pondered over. There are at least two flat contradictions to the general order of the atomic weights. Argon has an atomic weight slightly greater than that of potassium, and cannot do other than precede it. Whilst iodine has an atomic weight lower than tellurium, and cannot precede it in the classification without entirely breaking up similar families. These "exceptions" must be accepted as part of the scheme. In spite of all the determinations they have stimulated of the atomic weights of the elements in question, the anomalies have only been confirmed. After the first two periods, each of eight elements, the third begins and runs half through as far as Group IV before any difference becomes apparent. Then follow no less than ten members (vanadium to germanium) enclosed in the Table between brackets, which are, so to speak, interpolated suddenly into the period, before the characteristics we have been led to expect from the previous members of Groups V, VI, and VII fully reappear. The next period is a repetition of the last. There are five normal representatives of Groups 0 to IV, then nine "interpolated" elements (one is still unknown) exactly analogous to the preceding set of ten, before the elements of Groups V to VII are again met with. The next period begins normally with three perfect representatives of Groups 0 to II, and then a startling change takes place. When Group III, the group of the "earth elements," is reached, we begin the special family of elements known as the Rare-Earth Elements, of which no less than thirteen are known, with atomic weights fairly distributed between 139 and 174. They resemble one another with such extraordinary closeness, that the task of their separation and purification from one another is one of the most laborious and lengthy operations the chemist is acquainted with. For this reason it is difficult, even with the aid of the spectroscope, to say exactly how many exist. The members of this set of elements are all, like the earth-elements, trivalent. In them the simple principle of the Periodic Law, that each increase in atomic weight causes a step by step change in chemical character and valency, as illustrated by the passage from group to group, utterly breaks down. In spite of their name they now result in large quantities as a by-product of the thorium industry. With the exception of cerium, a minute proportion of which is used with thorium in the incandescent gas mantle, they are at present all technically worthless elements. Advancing in atomic weight beyond these elements we emerge at the beginning of a batch of nine new "interpolated" elements, perfectly repeating the characteristics of the last batch of nine, the same member being missing in each. After them only one representative of Groups V to VII (bismuth, Group V) is known, and in the next period the sole representatives are radium (Group II) and thorium (Group IV). The last element of all is uranium. These three elements, which end the Table, are the ones with the heaviest atoms known, and are all unstable, breaking up spontaneously and exhibiting in the process the recently discovered property of radioactivity.

What have been termed for purposes of

identification "interpolation elements" are all, without exception, well-defined metals. There are thirty places, and twenty-eight of them are known. They comprise some of the technically most valuable metals, including the iron group, and the whole of the "noble metals" used in the making of jewellery. They differ among themselves less markedly and distinctly in passing from member to member than in the other portions of the Periodic Table. At the same time the whole truth is by no means expressed by taking them entirely out of the Table and putting them in a part by themselves. The first three members have obviously resemblances of a most interesting kind with the Groups V to VII elements where they would naturally fall, and the last four with the Groups I to IV elements, especially in that their usual valency is the same as corresponds with these groups. For this reason it is usual only to take out the three middle members, the iron group and the light and heavy platinum metals, and to put these into the so-called Group VIII, as shown in the Table.

Nearly all the physical properties of the elements, their density, melting point, boiling points, etc., are, like the chemical nature, periodic properties — that is to say, they vary with the atomic weight periodically. Especially is this true for the density. Instead of density, or weight of unit volume, the relation is most clearly expressed by plotting against the atomic weight the atomic volume, which is the atomic weight divided by the density, and expresses the relative volume occupied by the individual atoms. The elements of Groups I to IV all occupy the maxima of the curve and the immediately following steeply descending portions. In this region increase of atomic weight is followed by decrease in the space occupied by the atom. The oxides of these elements show strong basic or alkaline character, uniting eagerly with acids. On the other hand, the characteristics of Groups V to VII are the opposite. Here the elements are on steeply ascending portions of the curve, increase of atomic weight causing great increase in the space occupied by the atom. The oxides of these elements are strong acids. What have been termed the "interpolated" elements occupy the minima between the peaks or maxima, and the atomic volume, like the other properties, does not show any great change in passing from one member to the next. The infusible and non-volatile elements occupy the minima of the curves throughout and in enumerating the elements which have been successfully employed in the manufacture of filaments for electric lamps, where these two qualities are essential, platinum, carbon, tungsten, tantalum, osmium (osram is a trade name for tungsten), it will be found that these elements all occupy low positions on the curve. Some of these elements, for example, tantalum, had never been prepared in the pure state prior to their technical production as lamp filaments.

The difficult task of evolving order and system out of the properties of the large number of distinct chemical elements, with their infinitely variegated natures, has thus been more or less satisfactorily accomplished. The Periodic Classification which connects all the properties with the one constant, the relative weight of the atom of the element, has proved a veritable mariner's chart and compass to the investigator. To the philosopher, however, it is a completely unsolved riddle, the meaning of which seems scarcely hidden beneath the surface and yet perpetually eludes the grasp. These numbers, expressing the relative weights of the atoms, are so many cryptic symbols which undoubtedly have a profound meaning, but which nevertheless remain mere numbers.

Hundreds of patient investigators have devoted years of strenuous labour, not a few their whole working lives, to the task of determining these natural constants with the greatest possible accuracy, and the work still progresses steadily. Some are known with a degree of precision unsurpassed anywhere in the exact sciences, and scarcely any are likely to be very seriously in error. The second or speculative method of attacking the problem of the nature of the elements has scarcely advanced far enough to deserve detailed consideration in a book of this character.

CHAPTER IV

HEAT AND THE KINETIC THEORY OF MATTER

SCIENCE is not an affair of watertight compartments. We cannot accept the idea of molecules and atoms in explanation of the properties of matter and deny their existence when we approach the study of heat. Like twin dewdrops forming side by side, these two great divisions of science have at length met and coalesced. The conception of the atomic and molecular structure of matter carries with it the complete and satisfying explanation of the kind of energy classed as heat.

Energy may sleep indefinitely in matter, in one of its numerous potential forms, without any indication of its presence. It is only perceptible when in the kinetic form as mechanical energy or the kinetic energy of moving masses, as electrical energy or the kinetic energy of moving electrons, and so on. Now in the real world as contrasted with the ideal world of the mathematical physicist, a convenient mental creation of frictionless machines, weightless moving parts, thermally impervious partitions, etc., energy when in the kinetic form invariably passes naturally into heat energy. If the kinetic energy is uncontrolled, the whole of it rapidly assumes the form of heat. By the use of properly constructed machines some part of it may be converted into other forms; for example, the kinetic energy of the electric current passing through an electric motor may give rise to a certain proportion of the total energy in the form of mechanical energy. In this case only a small part, represented by frictional and resistance losses, goes directly into heat. But in due course the whole of the energy sooner or later, and usually sooner rather than later, is quantitatively converted into heat. In the potential form, in coal, it has persisted for untold ages. Once released, heat is the sole ultimate product. By the law of conservation none is lost in the process, and the *total* amount of heat obtained is precisely the same, whether it raises steam in boilers, is converted into mechanical power by an engine, then into electrical power by a dynamo, is transmitted by conductors as electric current to a far distant place, there to be reconverted into mechanical power and back into heat again by friction. Heat energy may or may not be convertible into other forms, according to its temperature and the temperature of the surroundings. In the real world all substances conduct heat, and by this process all temperature differences are equalised naturally. The whole earth, with slight differences in different places maintained by meteorological causes, acts as the reservoir or sink of heat at uniform temperature, into which all the energy liberated in the world and assuming the kinetic form, flows without perceptibly affecting its temperature.

Heat when it has assumed the common level of temperature by natural agencies, such as conduction, is still energy in the kinetic form. Why does the process of transformation stop there? If energy is the universal commodity on which all life depends and all civilisation feeds, why has it not been found possible to draw on the immense reservoir of heat energy in the earth, the ocean, or the air, and to convert it back again, so solving, once for all, the dream of perpetual motion? Such a perpetual motion machine would not contradict the first law of energy, the law of conservation, but it would the second law, the law of availability. This law, realised as soon as the doctrine of energy and its transformation was developed, is, in origin, an arbitrary law, no reason being assigned. Like the Periodic System, it was part of a descriptive science, a statement of the facts found by experiment. Its philosophical foundations remained till recently as obscure as those of the Periodic Law.

Imagine a machine, installed, for example, on an Atlantic liner, capable of drawing upon the infinite supply of heat energy of uniform temperature in the ocean, cooling the ocean inappreciably and converting the heat into the mechanical work employed in propelling the vessel, the heat being returned to the ocean again by the process of friction between the vessel and the water. The second law states baldly that such a machine is impossible even to conceive, without any reference to impossibility of construction. This argument has been frequently used in thermo-dynamics, and, in spite of its apparently negative character, no theorem has been more fruitful in leading to positive discoveries. The favourite method was first to try to conceive such a machine. One of the earliest examples made use of the freezing of water. Water, unlike most other liquids, expands considerably in volume when it turns into ice, and, as is well known from the bursting of even strong steel vessels by the freezing of water, is capable of exerting considerable pressure, which, just as in the case of steam, can be conceived to be used to drive an engine. James Thomson found that a perpetual motion machine of the kind described could be imagined, fulfilling the scientific conditions rigidly, provided that the freezing point of water did not change when the pressure was increased. As befitted one of those associated with the development of the laws of energy, he did not imagine that he had found a case which upsets the second law, but he argued that the freezing point must change with increase of pressure, and, moreover, must become the lower the higher the pressure. Experiment proved that he was right. Ice melts under great pressure without rise of temperature, and freezes again when the pressure is relieved. Most other liquids which contract when they become solid have their solidification points increased by increase of pressure. Thus, by the use of the doctrine of the Impossibility of Perpetual Motion, there was predicted a quite unknown connection between the change of volume experienced by a liquid freezing on the one hand, and, on the other, the influence which increase of pressure has on its freezing point. This paved the way for a host of valuable predictions of a similar kind, connecting a variety of the physical properties of substances together, which by no simple process of reasoning could be shown to be at all connected. In no single case have these innumerable predictions been falsified by experiments, and they therefore afford a purely experimental proof of the doctrine denying the possibility of perpetual motion and of utilising the waste heat energy of uniform temperature.

To understand more clearly the reason why perpetual motion is impossible in the sense laid down, we have merely to open our eyes to see things as they are, and not as they appear to be to our grossly insensitive unaided sight. All matter is composed of molecules, or units of division of definite mass, and these almost infinitely small molecules do not occupy space continuously. Even if we could see these molecules individually we should not probably know any more about them in consequence than we do already; for of late years we have, thanks to radioactive phenomena, been able to apprehend them as individuals, and even to count their numbers with results in startling agreement with previous deductions. But if we could see matter as it is, instead of the gross masses which alone appeal to our senses, we should exclaim, "Perpetual motion impossible? Why, there is nothing else in the whole universe!" To do so we need not even imagine the impossible. All that is necessary is to look through a good microscope with a high power lens at some turbid liquid, that is, some liquid containing a cloud of very small solid suspended particles, just visible under the high magnification employed. We shall see not a single one of these particles at rest. Each is animated with the most lively and independent movement, darting hither and thither, turning and reversing all the time, so that the whole scene is a veritable dance of particles in and out of the field of vision. Hour after hour and year after year the dance goes on undiminished, independent of time and place and of the nature of the particles, except that the smaller they are the more lively is their movement. It is the Brownian movement, discovered in 1827 by the famous botanist Robert Brown; and this state of commotion, which is the normal condition of objects still large enough to be perceived, gives a faint indication of the normal condition of the almost infinitely smaller world of molecules.

Matter exists in three states, the gaseous, liquid, and solid. The condition of things obtaining in the first two of these states is very completely known, whilst of the third little that is quite definite can yet be said. The gaseous state is distinguished by the property of the matter, unless it is completely retained by a closed vessel, of diffusing out and filling all the available space offered to it uniformly. It is of no cousequence whether the space in question is entirely empty or vacuous, or whether it is already occupied by another gas. Each gas by diffusion fills the whole space uniformly as though the other was not present, so that after a time the composition throughout is uniform, whatever it was to start with. From this we may argue that when uniformity has been attained the process of mixing by no means ceases, though no further change of concentration in the different parts can occur. If a closed bottle of a gas with powerful odour is opened in one corner, rapidly the gas makes its presence known throughout the whole room; whilst if the bottle contained air only it would make its way throughout the room no less quickly, though no evidence of the fact might be apparent. The older school tried hard to account for these properties by assuming that the particles of gases repel one another. The present school found a more solid foundation to build upon when they regarded the molecules of a gas as in *perpetual motion.* So far from repelling one another, molecules in most cases tend to move together. When the kinetic energy of the moving molecules is reduced so much by lowering the temperature that this tendency can produce its effect, the gas condenses to a liquid.

The kinetic theory of gases, as it is called, imagines all the individual molecules of a gas to be moving about with very great velocities quite independently of one another, but incessantly colliding together with themselves and with the walls of the containing vessel, rebounding and incessantly repeating the process. These impacts on the walls generate the pressure that the gas exerts outwards, and if the walls were not there nothing, except the mutual encounters between the molecules, would hinder the free passage of the gas outward into space. The higher the temperature the greater the pressure exerted by any given quantity of gas confined in any space of given volume, and this increase of pressure with temperature is very nearly the same for all gases. Now, since the mass of the gas is not changed by the rise of temperature whilst the pressure is, rise of temperature must make the impacts on the walls either more numerous or more violent, or, in general, both more numerous and more violent. Hence it was deduced that equal rise of temperature increases the kinetic energy of all gases, regardless of the nature or mass of their molecules, to the same proportionate extent.

The problems that presented themselves to the mathematician in this theory did not prove as formidable as might have been expected. They asked themselves Avogadro's question. Consider two different gases, for example, hydrogen and oxygen. Let equal volumes of each gas, measured at the same pressure and the same temperature, be compared. What are the relative numbers of molecules in each? Avogadro, it will be recalled, had answered this question simply by intuition, and had stated that the number of molecules in equal volumes of all gases measured under the same conditions is the same, and thus had laid down the fundamental generalisation which had enabled chemists to determine relative atomic weights and to discover the Periodic Law. The mathematicians approached the subject from the view that since the gases are (1) at the same pressure, the kinetic energy of the molecules taken as a whole in each set must be the same, and (2) at the same temperature, the kinetic energies of each set of molecules will remain the same as they were before, after they are mixed. In investigating the general conditions for two sets of independently moving molecules to be able to intermingle without any transference of energy from one set to the other, they deduced that, on the average, the kinetic energy of the *individual* molecules in the two sets must also be equal. But the total kinetic energy is the same for both gases. Therefore the numbers of their molecules must be equal. Avogadro's rule became merely one consequence of a great general dynamical theorem which applies to *any* system, whether a gas or not, which consists of a vast swarm of free unhampered and independently moving small units of mass, incessantly colliding with each other and rebounding. The theorem is known as the law of equipartition of energy. The molecule of hydrogen, on the average, has the same kinetic energy as the molecule of oxygen at the same temperature. Since the kinetic energy is proportional to the product of the mass into the square of the velocity, and the molecule of oxygen is sixteen times as heavy as that of hydrogen, the mean velocity of the hydrogen molecule must be four times that of the oxygen molecule, and so on for all other molecules. Gradually all uncertainty has been eliminated from this conception; and, as already remarked, it would add little, probably, to our knowledge if we could weigh, count, and measure the velocity of these molecules of gases individually. Every cubic centimetre of any gas, measured under standard conditions (0° C. and 760 millimetres, barometric pressure) contains twenty-seven million million million molecules. The weight of the single molecule of hydrogen is about three million-million-million-millionths of a gram, and its velocity at 0° C. is rather more than a mile a second. The hydrogen molecule is, it is true, the smallest and simplest molecule of matter known, but it is a large and sluggishly moving individual compared with another known particle, the electron or atom of negative electricity. The average kinetic energy of all molecules is increased to the same extent by rise of temperature, as also is necessary from the law of equipartition of energy, which does not hold merely for one particular temperature, but for all temperatures.

Now what, exactly, is meant by temperature? The various thermometric scales of Fahrenheit, Centigrade, and Reaumur came into use before the nature of heat was fully understood, and all started from the properties of the substance water. Thus the Centigrade scale, adopted in science, calls the freezing point of water zero, or 0° C., and the boiling point of water under normal barometric pressure 100° C., the interval between being divided equally into 100 parts or degrees Centigrade. Temperatures below zero Centigrade are represented by the negative sign. Now it is quite obvious that if heat energy is the kinetic energy of moving molecules, although there is no necessary upper limit to the extent a substance may be heated and to the temperature it may acquire, there is a very definite limit to the extent that it can be cooled. A substance composed of molecules at rest is absolutely cold, and no substance can be imagined to be colder. The absolute zero of temperature is the true zero of a thermometric scale, not the freezing point of water or of any other substance.

Few people who have not thought deeply in the subject realise what a confused medley of vague and disconnected ideas are concealed within the common uses of the word *Temperature*. In truth, the term, even in science, has changed its significance almost with every decade of the last century. It is more convenient to defer the discussion of what we really mean by the word "Temperature" until the next chapter. It may merely be remarked in passing that the term in its philosophical sense has a far more fundamental meaning than that used in the science of practical thermometry, which refers temperatures to the instruments and scales in current use, just as weights and measures are referred to the arbitrary standards set up and preserved by the various standardising institutions. The kinetic theory of gases and the law of the equipartition of energy give a perfectly definite meaning to the word, and fixes one rational thermometric scale, known as the thermodynamic or absolute scale, which does not differ seriously from the earlier thermometric scales on the one hand or on the other from what might be regarded as the natural meaning of the word *temperature*, or intensity of heat, impossible as that is yet to define. This truly remarkable coincidence of the old thermometric scales with the thermodynamic scale, and with the true scale, when we are in a position to define it, has had, of course, great practical advantages. But they have been more than counterbalanced by the confusion on the philosophical side that has resulted by the use of a term, as a strictly scientific term, long before it had a single unchanging meaning.

The scale we are about to set up, the thermodynamic scale, is in a sense an arbitrary scale like its predecessors. Its connection with the natural scale must be left over for future discussion. Its basis is the law of equipartition of energy. Two substances have the same temperature as that of a gas —and therefore according to Euclid's axiom are themselves of the same temperature— when they can be brought into thermal communication with this gas without altering the kinetic energy of translation of its molecules, that is, without altering its pressure when its volume is kept constant or its volume when its pressure is kept constant.

This basis, which rests on Davy's neglected generalisation in 1812, that the laws of communication of heat are precisely the same as the laws of communication of motion, affords at once a scale of temperature. That temperature will be twice another, at which the kinetic energy of translation of the molecules, or, the pressure at constant volume, of a gas is doubled. If we adopt the Centigrade scale we find that the pressure of a constant volume gas thermometer increases for each increase of 1° C. by 1/273rd of the pressure of the gas at 0° C., and decreases by the same amount for each 1° C. below 0° C.

In consequence, at a temperature of 273° C. the pressure of the gas is doubled, at 546° C. it is trebled, whilst at 819° C. it is quadrupled, and so on without limit so far as is known. Conversely, coming down the scale, the temperature will be halved at $-136.5°$ C., whilst at $-273°$ C., if the gas did not liquefy, as all known gases do before this temperature is reached, the pressure and kinetic energy of the molecules would be reduced to nothing. The molecules would be absolutely still and therefore absolutely cold, and nothing can be conceived to be colder. In other words, the absolute zero of temperature is 273° below zero on the Centigrade scale. The Absolute Scale of Temperature, as it is called, is thus obtained by adding 273° to the temperature expressed in degrees Centigrade.

It will be observed that there is nothing inferred here of the intensity of heat, that is, the quantity of heat energy per unit quantity of substance which is implied in the natural definition. If we tried to follow such an idea at present we should be landed in a quagmire of difficulties, sufficiently indicated by the older distinction of two kinds of "heat," "sensible heat" and "latent heat." Still less is the thermodynamic scale referred to gases because they are practically perfect or ideal thermometric substances. Such a substance would be increased in temperature by twice the amount by the communication of two units of heat energy as it would be by the communication of one unit of heat energy. This would only necessarily be true if all the heat was employed in increasing the kinetic energy of translation of the molecules, that is, in raising the temperature. In this sense the common gases are very far from fulfilling the condition laid down. We have merely stretched a tight-rope across an abyss of uncertainty to gain entrance into regions beyond, because the progress of science is rapidly converting that tight-rope into a bridge, with handrail complete for the most timid adventurer.

The extension of the kinetic theory to liquids was one of the great advances of the closing decades of the last century. Liquids, we have seen; result by condensation from gases when the mutual tendency to draw together, which the molecules exhibit, becomes, by the lowering of the temperature, able to control the kinetic energy of the molecules and to restrict their movements so that they are no longer able to wander freely away into space. The essential differences between the liquid and gaseous states are, however, confined solely to an exceedingly thin skin at the boundary or surface of the liquid. *Inside* the liquid an unhampered independent motion of the molecules, moving by virtue of their own kinetic energy, and not restricted in any way except by their incessant mutual encounters, is going on exactly the same as in gases. The molecules are, it is true, generally much more closely packed in liquids than in gases, and in consequence they collide much more frequently. The "free path," as it is called, in liquids, or the average linear distance each molecule moves before it experiences an encounter with another, is exceedingly minute. Even in gases under standard conditions this free path is, as a rule, a distance below the microscopically visible. In consequence, in liquids the molecules move in very zigzag paths, being perpetually turned back the way they came, and having to traverse great distances without moving very far away from their starting-point, so that in spite of their great velocities they do not diffuse very rapidly. It might be supposed that, owing to the natural tendency of the molecules to draw together, or in the phraseology of the day to their great mutual attraction for one another, the character of the motion in liquids must be different from that in gases. This is not the case *inside* the surface skin. In the interior the so-called attractions are uniform in all directions and therefore cancel out. The molecule is attracted equally in all directions at once, and hence its motion is precisely the same whether such attractions are supposed to exist or not. It is only at the surface that the state of liquids is essentially different from that of gases. For at the surface the tendency of the molecules to move together, that is, inward, is quite unbalanced by any compensating tendency to move outward. This causes at the surface a very strong tendency for the molecules to move towards the interior of the liquid, which is called Surface Tension, and to which practically all the properties of liquids, in so far as they are distinguished from gases, are due. Inside all is freedom of movement, a mad rushing hither and thither of an altogether disconnected swarm of molecules moving in every direction at once, colliding and rebounding, but never stopping. In liquids the crowd is much more closely packed, and the motion, therefore, is even more chaotic than in gases; but otherwise there is no difference. Now the law of equipartition of energy states that in *any* such system of free unhampered and independently moving units, incessantly colliding with each other, each separate molecule, nay more, each independent particle, whatever its mass, on the average must possess the same amount of kinetic energy. This amount of kinetic energy at 0° C. is precisely that possessed by a molecule of hydrogen, weighing 3 million-million-million-millionth of a gram moving at a speed of rather more than a mile a second. If the particle weighed a milligram, the total distance it would move in a second, even if its almost infinitely entangled zigzag path were straightened out, would be scarcely visible under the microscope, so that we must not expect weighable quantities of matter immersed in a fluid to move about visibly. But what if we deal with the smallest particles it is possible to see under the microscope? Then we shall see what Robert Brown, all unaware of its true and tremendous significance, first saw with mortal eyes, the perpetual motion of the molecules, which mathematical physics, following Davy's dictum and regarding molecules as freely moving masses had, by means of the embracing theorem of the equipartition of energy, arrived at with the eye of faith.

Of course, in the Brownian movement we only see the effects of the actual mad rush of the molecules themselves. The smallest particle visible under the microscope contains millions of separate molecules. If we begin by a particle visible to the naked eye suspended in a fluid, on account of its relatively large size, it is subjected to a rain of impacts from the molecules which is practically equal on all sides at the same time and the effects of which cancel one another out, so that the particle remains practically at rest. But if we reduce its size gradually, so making it at the same time lighter and quicker to respond to molecular impacts, we shall arrive at a point at which the molecular impacts do not always so cancel out. At one instant more molecules may hit one side than the opposite, and the light particle will instantly dart off in the appropriate direction. Before it has moved an appreciable distance, some other combination of simultaneous impacts will have occurred and the direction will change, and, in the end, the particle, if only small enough, will never be for a moment still, but will be agitated perpetually by the bombardment to which it is on all sides subjected by the invisible swarm of molecules.

Some beautiful experiments of M. Perrin have recently subjected this Brownian movement to quantitative investigation. By a simple ingenious process of fractional centrifugalisation, he succeeded in preparing emulsions of gamboge and gum mastic in water, the particles of which were all of the same size, which could be measured under the microscope. From the size and density of the substance the weight of the particle could be deduced. By the aid of the researches of numerous mathematical physicists he was able to calculate the mean kinetic energy of translation from the results of his observations of the movement under the microscope. He examined particles varying in mass over a range of 60,000 to 1 and found that the mean kinetic energy was independent of the mass, and was, moreover, the same as that of a molecule of hydrogen or any other gas at the same temperature. The largest particles he examined were just visible with a simple lens in sunlight, and weighed 100,000,000,000 times as much as the molecule of hydrogen, and the movement of such relatively large particles is small and sluggish. With the smallest visible particles the motion, or rather commotion, is extraordinarily lively. What then must be the character of the invisible turmoil going on in the actual molecules of every drop of liquid and every part of the atmosphere around us?

Perpetual motion, proportional in magnitude to the square root of the absolute temperature of the substance, is the universal condition of the liquid and the gaseous states. As regards the solid state far less is known. It seems clear that any freedom of translatory motion is an impossibility for a crystalline solid, for the crystal form is the outcome of the molecules occupying definite geometrical positions in space with reference to each other. But vibratory motion in constrained paths there must be among the molecules of a solid, increasing with the temperature until the molecules drag their anchors, as it were, and the substance melts.

What inexhaustible energy—the waste energy of heat of uniform temperature—these molecules everywhere around us possess! But the second law of the Doctrine of Energy states that this source of energy must for ever remain useless for us. Still the mind is not satisfied, and again it asks for the reason.

Let us consider the process of the conversion of the energy of mechanical motion into the energy of heat, by resistance or friction, and, to take a simple case, suppose that water is spinning round or rotating in a pail. A cork floating on the surface or a ball of the right specific gravity floating below the surface moves with the water as it rotates. Its motion becomes slower and slower and finally ceases. At first, superimposed on the mad rush of the molecules in every conceivable direction at once, was a directed motion of the mass as a whole, each portion having a definite resultant motion in a circular path around the centre of the pail. At the end, when all is still, we know that although the cork or floating ball remains still, the mad motion of the molecules endures. A suspended particle sufficiently small will show the Brownian movement anywhere in the liquid. The mechanical energy of the originally moving water has, we know, been frittered down into heat, with the consequence that the temperature and hence the separate motions of the individual molecules has been increased to an infinitesimal extent. In other words, the directed or co-ordinated motion of the mass as a whole has been converted by the process into the perfectly mad or decoordinated motion of the individual molecules. The motion and the energy of the motion has not been destroyed. It has merely been divided up amongst the smallest particles, all sense of the original direction being lost and a perfectly uniform distributed motion of all the particles in all possible directions having taken its place. The conversion of mechanical energy into heat by friction involves only loss of direction and of co-ordination. The kind of perpetual motion which is not only possible, but which, indeed, is the only universal condition of matter, is the perfectly useless decoordinated motion of swarms of molecules. The kind of perpetual motion which is impossible in the real, as contrasted with the "ideal" world, is the motion which retains its dirccion unimpaired. The problem of achieving perpetual motion contrary to the second law is that of bringing order and direction once more into the chaotic rush of the molecules, to marshal and drill the mob so that once more they can act together to produce a common effect.

If an insight has been obtained into the general nature of the Second Law of Thermodynamics, its quantitative aspect, which is of the utmost practical importance in engineering science, will not prove altogether unintelligible. Heat at the uniform temperature of the surroundings cannot practically be transformed into mechanical work, although theoretically this might be accomplished by the operation of an intelligence directing and regulating the traffic, as it were, among the individual molecules. If heat were so transformed, the substances in which it was contained would be cooled to a lower temperature than before to an extent directly dependent upon the amount transformed. The second law states that such spontaneous cooling below the temperature of the surroundings is not possible. The various refrigerating machines which practically effect cooling do so paradoxically by converting mechanical work into heat, and later a simple case of such a cooling will be considered.

Spontaneous cooling of a substance can only occur continuously without the performance of work when the surroundings are at a lower temperature than the substance. This cooling may occur in a perfectly natural manner by means of simple conduction of heat, in which case all of the heat energy given out by the hot substance still remains as heat energy in the surroundings at the lower temperature; or the fall of the temperature of the working substance may occur in an artificial or directed manner through an engine which transforms a part of its heat energy into work, so that only a portion of the heat, the untransformed portion, remains in existence after the cooling has occurred to lower the temperature. It is possible to direct and regulate the flow of heat among masses of matter but not among molecules. The mechanism of the conversion will be considered in the next chapter. If we try to imagine a complete conversion into work of the whole of the heat energy contained in a certain quantity of substance, we see at once that thereby the temperature of that substance must be reduced to the absolute zero, −273° C. The process of cooling, however, stops naturally when the temperature reached becomes that of the surroundings. Thus, if we consider a simple substance cooling without any change of state, the heat it will evolve in cooling any number of degrees is only a part of its total heat energy, and this part is calculable. If it cools from 100° to 0° C., this is from 373° to 273° absolute, the proportion of the heat energy remaining in the substance after cooling to that initially present is as 273 to 373, so that the proportion that has left the substance is 100/373 of the amount initially present, or 100/273 of the amount finally present. This fraction 100/373 of the initial heat, in the example chosen, is all that leaves the substance, and therefore is the maximum available for conversion into work. So that if a perfect transformation of this portion into work were possible, the efficiency of the process would still be only about 27%.

Whereas of an electric motor an efficiency of 90 to 95% is expected, that is, it will transform this fraction of the electrical energy supplied to it into mechanical work under proper conditions. We can increase the efficiency of the transformation of heat into work in only two ways. The first is to lower the final temperature attained, which is impossible, practically, as the final temperature is fixed as the prevailing temperature of the earth and air. The second is to increase the initial temperature. Obviously the higher the initial temperature the greater the proportion of heat available for transformation. For this reason gas and oil engines, which work between wider limits of temperature than the steam engine, are thermodynamically much more efficient. It might be thought that these simple considerations, so easily understood when one deals with a substance cooling without change of state, might not apply to the case where, as in the steam engine, changes of state from liquid to vapour and from vapour back into liquid occur. Experience proves that they do apply.

The relation already deduced between the initial and final temperatures and the maximum possible efficiency of the conversion process is of perfectly general applicability, and so far as is known is universally true of all artificial heat engines. If T_1 is the absolute temperature of the source of heat, or boiler, and T_2 that of the reservoir of heat, or condensor, the possible efficiency of the process, or the ratio between the quantity of work produced and the quantity of heat supplied, is T_1-T_2/T_1. It is hardly necessary to say that this maximum theoretical efficiency is not even approximated to in any actual engine, not through any fault of the engineers, but because the ideal theoretical conditions are entirely different from, and in many cases diametrically opposed to, those obtaining in practice.

The question may now be asked whether the limitation expressed by the second law is necessarily and permanently true, or whether it is merely an expression of the fact that, so far, no process has been discovered competent to bring order and direction into the chaotic rush of molecules. As has always been understood by the best exponents of the law, the existence of any sort of "molecular intelligence" would vitiate the second law. For though, throughout, we have spoken of mean velocity, mean kinetic energy, etc., as the statistical averages applying to the whole immense swarm of molecules, averages which have perfectly definite values, the individual molecules in the same mass of matter depart from these average values widely. In a gas at uniform temperature some small proportion of the whole are moving with speeds many times greater than the average speed, and some small proportion at speeds many times less than the average. Even if all the molecules began at the same speed, the incessant collisions among them would at once result in inequalities. There is a perfectly definite distribution of velocity, and the fraction of the total number of molecules actually possessing the mean velocity, or any assigned velocity above or below this mean, can be calculated perfectly satisfactorily. Hence in a gas of uniform temperature there are, from a molecular point of view, wide variations of temperature, and there seems no reason why, for example, molecules should not in some cases be able to utilise this, to them, available energy and convert it into other forms available for the larger world.

The question has often been mooted, for example, whether the processes of life obey the second law. Certainly from some points of view a horse or other beast of burden is an extraordinarily efficient machine, as Count Rumford perfectly well knew. How efficient, possibly was not generally realised until motor cars came in, and instead of two horses to pull the car it was found, in the absence of a whip, that twelve or fifteen was an appropriate minimum number, or the car would stick on every moderate hill. The so-called horse-power, which furnishes the engineer's unit, is not the power of an ordinary horse either, but was fixed by tests on the best British cart horses which could be procured. Have the minute cells of the body the power of taking advantage of the difference in the temperature of the molecules bombarding them, and when one comes along at more than average speed, absorbing it and its energy, building up a larger cell thereby, which in course of time undergoes metabolism and evolves again its store of energy? It is a fascinating and a legitimate line of inquiry, but time alone and experiment will answer it decisively. With the full understanding that the second law merely states that it is impossible usefully to direct the chaotic movements of molecules or to co-ordinate them, respect for it as one of the necessary fundamental laws of science will be weakened and applications of it to entirely unknown phenomena only made with caution.

Particularly where intelligence operates is this the case. Life in its lowliest is a single cell, vastly complex, it is true, from the molecular point of view, but still small enough to be agitated by the Brownian movement. Brown actually thought, when he discovered the phenomenon called by his name, that the particles he was observing were alive, so closely does their motion simulate that of the smallest organisms. If life begins with a single cell, does intelligence? Does the physical distinction between living and dead matter begin in the jostling molecular crowd? Inanimate molecules in all their movements obey the law of probability, the law which governs the successive falls of a true die. In the presence of a rudimentary intelligence do they still follow the law, or do they now obey another law, the law of a die that is loaded? If so, mathematicians will have to extend their investigations to cases where the laws of simple probability do not act alone, and their domain to fields into which they have not hitherto entered. For the living body is, after all, a machine first. We know machines that are without life, but we do not know life apart from its machine. The modern world, which has recently learned that it is useless to try to educate a child that is not fed, has had its attention directed somewhat forcibly to this aspect of life.

The suggestion just made is, however, but one of many views worth discussing. Another is that life processes do not obey the second law, for there seems to be a consensus of opinion that they cannot possibly obey it, because the chemical energy of food suffers direct transformation into work without first being converted into heat. As will later be discussed, the second law would not then operate. Now this is just another problem which is of immense practical importance. Many chemical changes can be made to take place in such a way as to yield their energy not as heat, but as electrical energy directly. But such a result for the most important of all chemical changes, that equivalent to the combustion of fuel, has not so far been achieved. Be the explanation what it may, the living body is an extraordinarily efficient machine, not so much on the side of the absolute value of its efficiency, of which, indeed, far too little is known, but because it deals with kinds of chemical change which cannot be converted by inanimate engines into useful forms of energy without terrible waste.

CHAPTER V

POTENTIAL AND CHEMICAL ENERGY

It is curious to reflect upon the reason why mechanics was an exact science from the time of Newton for nearly two centuries before the doctrine of the conservation of energy was established. Newton in his first law of motion stated an hitherto unrecognised great truth. The natural and simplest kind of motion is uniform continuous motion. This simple uniform continuous motion, unchanged in amount or direction, is never realised practically except among the stars. In no other case is motion entirely free and unhampered. On the earth, friction and the resistance offered by the air gradually reduce a moving body to the state of rest. The velocities of the planets, whilst they remain sensibly uniform in amount, change their direction unceasingly and follow circular rather than straight paths, because the planets are not alone in space, but are in proximity to a much larger sun, towards which their motion tends. In empty space a mass of matter, if moving initially, moves on unchanged in direction or speed for ever and ever, and Newton was the first to recognise it. In dealing with actual cases of motion in the heavens and earth, Newton fell into the common error of his day. He imagined causes to exist for the departure of these motions from the natural or simple law, and it has taken science three centuries to recognise that the causes imagined are not real causes, and that they only describe the effects, without any more light upon the origin of these effects. The imaginary cause of change of motion, Newton defined as Force. Great men have more to fear from their followers than from their opponents, for Newton's mind was such that it is impossible that he himself could have remained long under any misunderstanding as to what Force really was. The universal attribute of matter, whereby it gravitates, was accounted for by imagining that all matter attracted all other matter with a force which was termed the force of gravity. The law of gravitation stated that the force of gravity varies according to the product of the masses of the matter gravitating, and inversely according to the square of their distance apart. No exception can be taken to this view as long as it is recognised clearly that it expresses simply that the motions which matter executes in space *are what they would be* supposing the force of gravity obeying the law stated did exist. But whenever the motion of matter departs from the simple law, appropriate forces have been imagined to exist. Thus in the condensation together of the molecules of a gas to form a liquid the molecules are imagined to attract one another. The property of the electric charge of spontaneously dissipating itself so that it always resides on the surface of a charged body, is spoken of as being due to the repulsion of similar electric charges. Atoms of matter are regarded as attracting or repelling one another with the force of chemical affinity, a phrase which makes every thinking man shudder. So has grown up the preposterous notion that forces really exist and are the permanent attributes of masses of matter, molecules, atoms, electricity, etc. When a weight rests upon the table the force of gravity is supposed to be "acting" all the time. The earth is regarded as attracting it with a constant pull, and the "reaction" of the table prevents the weight falling. This is not the proper view to take. It is true that in the science of mechanics, where gravitation is the dominating phenomenon, it is convenient. But the conception of force and its pseudo-physical reality undoubtedly delayed for centuries the recognition of the law of the conservation of energy. Only what is conserved has the right to be considered a physical existence. Long and mistaken were the attempts to arrive at a conservation of forces. In other branches of science the conception of force is a stumbling-block and a delusion. The purest example of motion according to the first law is the motion at constant temperature of molecules of a gas taken as a whole. It is true that the direction of motion is unceasingly changing by impacts of the molecules upon the walls of the containing vessel and by mutual collisions between the molecules. Interchange of motion, and therefore of the kinetic energy of motion, is always going on. But considered statistically, that is, keeping in mind only the average motion of the molecules, or their total kinetic energy, no change results as a consequence of these countless collisions. Thousands of millions of times a second for each molecule encounters occur, reversing or altering the direction of the molecules. According to the definition of "force," intermittent "forces" come into play thousands of millions of times a second and disappear, just as in the collision of billiard balls. Whereas, since two molecule or billiard balls cannot occupy the same space at the same time, it is more natural in such cases not to invent "forces" between the molecules to account for their departure, as individuals, from the law of simple uniform motion, so making them appear to exert forces rather than to pursue the easiest paths available. It is sufficient simply to regard the twists and turns of their motion as due to their relative *position.*

Then the truth emerges that all so-called forces are positional. Whether anything "attracts" or "repels" another is a question not only of the nature of the "attractor" and "attracted," but also of their relative position. The fact that by no means can we put ourselves into a position not dominated by the effects of gravitation, led erroneously to the idea that forces still act even when they produce no effects. It is contrary to the spirit of modern science to imagine the existence of causes which in reality are names only, and which throw no light at all on the effects they are supposed to explain. Yet we still speak of things attracting, repelling and exerting forces as if they were actual agents. We—in our anthropomorphic fashion—invest them with human attributes, just as the Greeks invested their gods.

The reason is that the English language is quite unable to express motion, other than that according to the first law, except by words implying cause. We have to *say* two bodies attract and repel one another when we *mean* that they naturally tend to approach or to move away from one another, and if no obstacle stands in the way, do so move. The words "attract" and "repel" have no intransitive forms to express the effects of which they themselves signify the cause. Unless these terms are coined, the language fails to express the ideas of science. There is one word, "to gravitate," which simply expresses motion of a particular kind without implying any idea of cause. So let us invent the terms "to tractate" and "to pellate" when otherwise we should be forced to employ "attract" or "repel," and speak of bodies "tractating" and "pellating" when otherwise it would be necessary to say that one body attracts or repels the other.

When the tractation or pellation is over,

no imaginary forces remain to bother about, for these terms express processes rather than their unknown causes. But these natural processes may be reversed. Bodies which tractate may be made to move farther apart, and this must therefore be termed "detractation," the moving apart of bodies which naturally approach. Similarly we have "depellation," the moving together of bodies which naturally move apart. So equipped, we may now hope to consider some phenomena where motion is not the simple motion according to Newton's first law, but depends upon the position of the moving bodies relatively to one another. In other words, we pass from the consideration of kinetic energy to that of potential energy. In tractation or pellation, bodies spontaneously draw together or draw apart, even though at rest initially, and acquire therefore kinetic energy. In each process the appearance of kinetic energy accompanies the change of position. Hence the bodies in their original position, though possibly possessing no kinetic energy, have energy—if energy is real and not a delusion—which is associated with their position. This is what is meant by potential energy. Tractation and pellation may therefore be defined both as processes in which energy of position is converted into energy of motion, in the first case, by the shrinkage or concentration, in the second case, by the expansion of the parts of the system.

It may be said at once that potential energy is a way of expressing facts, not supposed explanations of them. Why two bodies tractate or pellate is not known in a single instance, least of all perhaps in the oldest recognised case, gravitation. An ingenious theory of gravitation was put forward a century ago which, though not accepted, is very suggestive, and illustrates the difference between what science would consider a real cause and one that is fictitious, like the "force of gravity." Le Sage imagined that all space was filled with "ultra-mundane corpuscles," or particles smaller than material particles, flying about in all directions, moving in straight lines according to the first law of motion, and possessing kinetic energy. A single world in space is subjected to the rain of these corpuscles from all dircetions at once, and so remains still. Two worlds in space both are similarly bombarded, but each screens the other from the particles coming in a particular direction, and so they are urged towards each other! Such explains beautifully one part of the law of gravitation, that the effect decreases as the square of the distance apart of the bodies. But it would be natural to suppose that the area rather than the mass of the body would condition its screening effect.

Le Sage and Newton both "imagined," but the former's hypothesis is no verbal restatement of the effect. It traces gravitation to imaginary corpuscles, but, if it is correct, these corpuscles are real. So long as gravitation is the only effect these corpuscles produce, and so long as they remain unknowable apart from gravitation, the two explanations are on a par. But if such corpuscles are real they should ultimately, as science advances, become knowable, and hence the truth of the explanation can be examined. In this sense it is an attempt to find a real cause. Science of the last decade has moved very near Le Sage's conceptions. We know rays—the γ-rays of radium—which penetrate several inches of lead before they are stopped, and which are supposed by some to be corpuscles or particles smaller than atoms. But the difficulties as well as the possibilities of this view of gravitation have also been thereby increased. The value of the hypothesis is that it gives us in one case, if we accepted it, an idea of what potential energy might be, namely, the kinetic energy of what has so far remained outside the ken of science.

Frankly it must be admitted, however, that the real causes of tractation and pellation are unknown, though not necessarily unknowable. That being so, we accept the observed phenomena as facts, and find out all we can as to how they operate, and in this field much is known. The case of the tractation of steam into water and of the pellation of water into ice has already been referred to. When the molecules of the gas, steam, are compressed or crowded together to an extent which must be the greater the higher the temperature, the steam condenses into water without change of temperature, but a large amount of heat, the so-called latent heat of condensation, is produced. If this is not allowed to escape, the condensation stops. The heat generated will not raise the temperature of the steam, even if it is not got rid of. Neither, when the heat is allowed to escape, does the temperature of the steam fall, so long as any remains uncondensed. Conversely when steam is raised from water, at a temperature which must be the higher the greater the pressure, the molecules of the liquid are changed into molecules of the gas, and a great amount of heat—the latent heat of vaporisation—is absorbed without increasing the temperature. The difference between tractation and pellation and simple concentration or expansion is that kinetic energy appears invariably as the result of the change of position. A perfect gas, and the same is nearly true of the common gases, neither tractates nor pellates. Whereas condensation of a gas to a liquid is invariably a tractation, whilst the formation of a gas from a liquid is a detractation. In the first kinetic energy appears as the so-called latent heat of condensation, in the second kinetic energy disappears, and this disappearance is termed the latent heat of vaporisation. This confusing term, "latent heat," does not express heat at all, but the change of energy of motion (heat) into energy of position, or *vice versa.* The heat energy of the molecules is the same whether the molecules are liquid or gaseous, but the latter have, in addition, energy of position. When they condense to form a liquid, kinetic energy is evolved. In the freezing of water into ice the molecules pellate and acquire kinetic energy (latent heat of liquefaction), which must be removed before the freezing can continue. In the melting of ice into water the molecules depellate, and kinetic energy or heat disappears (latent heat of fusion). These simple cases show that a pellation is not merely the opposite of tractation, *i.e.* a detractation. If the freezing of water were a detractation, kinetic energy or heat would be absorbed instead of being set free. Water, however, is rather exceptional, and the usual behaviour is for molecules to tractate when they pass from the liquid to the solid phase, just as they do when they pass from the gaseous to the liquid phase. Why they do so may be imagined but not demonstrated.

As explained in the Volume in this Library on Astronomy (p. 40), the tractation or gravitation of the matter when a sun shrinks may convert enormous quantities of energy of position into energy of motion or heat. In the tractation of masses kinetic energy appears.

In the tractation of molecules, heat results. No machine is known capable of employing the tractation of the molecules to produce mechanical energy directly. Hence all heat machines, whether they depend on tractation or not, obey the second law. The kinetic energy appears first as an increase in the uncoordinated movements of the molecules, and can only be partially coordinated for any common purpose.

Having considered, so far, what the second law teaches, it will be instructive to refer to a common error. The law certainly does not state that none of the heat energy of uniform temperature can under any circumstances be converted into mechanical energy. A clear picture can be obtained as to the actual mechanism in any heat engine by which the heat is converted into work.

In any cylinder and piston filled with a gas or vapour, the molecules impinge on the fixed walls of the cylinder and on the stationary piston and rebound without loss of energy like the perfectly elastic particles they are. But if now the piston, under the impacts of the gas molecules, commences to move outward, carrying with it all the connected machinery, part of the thermal energy of the flying molecules is converted into mechanical energy, and the gas is cooled. The molecules striking the receding piston do not rebound with their initial energy as they do when they strike the stationary wall of the cylinder, any more than a tennis ball received by the player on a receding racquet rebounds with the same energy as if the racquet were held rigidly. Conversely if the piston is entering the cylinder, compressing the gas, the molecules striking it rebound with more than their initial velocity, and the gas is heated, the energy of the moving machinery now being spent in heating the gas. The tennis player, on the analogy, now advances his racquet to meet the ball, driving it back with more than the energy with which it was approaching him.

If a bottle of compressed gas is opened, the gas rushes out into the air and does work making room for itself and forcing the surrounding atmosphere out of the way, or it may be used to force out the piston from a cylinder. It cools at once, and the bottle may become so cold that the moisture of the air condenses upon it as snow. Here, then, is a case where we start with everything at the uniform temperature of the surroundings, and convert part of the heat into mechanical work, cooling the substance below the temperature of the surroundings. If we had done no more than this we might be inclined to think that the second law could not be true. As a matter of fact the jinnee of the bottle has been allowed to escape. We have let the gas expand, and to compress it back to its original volume the process is exactly reversed, the work before gained is all spent and converted back into heat.

The principle of all refrigerating machines

will now be clear. Air or other gas or vapour at high pressure is allowed to expand, so doing work and becoming cold, cooling the substance it is desired to freeze. But air at high pressure is not a natural commodity, and so an engine is first employed to compress the air, doing work on it, which work is turned into heat and the air becomes hot. This heat is removed by circulating the compressed air through cooling tubes immersed in flowing water, and the cooled compressed air, when now allowed to expand, becomes as much colder than its surroundings as it became hotter when compressed. Paradoxically, refrigeration is effected by converting mechanical energy into heat, in order to be able to reconvert heat back into mechanical energy at a later stage, when the working substance has been cooled to the temperature of the surroundings. The second law which applies to the whole process is therefore borne out in this case.

Imagine a tall tank filled with water reaching from the lowest hold to the uppermost deck of a liner. Opening a hole in the bottom of the vessel beneath the tank will effect the discharge of its contents, until the water-level in the tank is the same as that of the surrounding ocean. Or, if the tank were empty to start with, the opening of the hole would cause the tank to fill to the same level as before. To empty the tank completely or to fill it completely requires pumping. The cooling of a hot substance or the heating of a cold one are natural processes which occur spontaneously like the partial emptying or the partial filling of the tank considered. The heating of a substance above the temperature of its surroundings or its cooling below this temperature are not natural spontaneous processes, but are only to be effected by artificial processes analogous to pumping. The whole matter is summed up in the statement that we are living in an immense ocean of heat energy at constant level, and efforts to raise heat above or depress it below this level are no more to be thought of without the expenditure of work than the elevation or depression of the level of the sea. Davy's dictum that the laws of communication of heat are precisely the same as the laws of communication of motion is still the best general expression of the reasons underlying the second law.

This commentary upon the molecular aspect of heat would be unprofitable if it did not enable certain features in the all-important processes of thermodynamics to be seen more clearly than the more formal presentations allow. By the law of the equipartition of energy we obtained a definite conception of temperature—at least of temperatures referred to gases. Temperatures on this scale are proportional to the average kinetic energy of translation of freely-moving molecules. There is, however, an important limitation, because no gas absolutely obeys the mathematical conception of a gas, or, in other words, actual gases depart under some conditions widely, under other conditions extremely slightly, from the *ideal* gas. The molecules of an ideal gas are points occupying no volume, and neither tractate nor pellate when the gas contracts or expands, that is to say they are really free-flying and independent. Hydrogen and helium under ordinary conditions approach so nearly the condition of ideal gases that scarcely any error is involved in their use as thermometric substances. The helium or hydrogen constant volume gas thermometer, in which the pressure of the gas is used to measure the temperature is the standard to which all others are referred. These are the two last gases to liquefy when temperature is reduced. Their readings agree with one another to −210° C., that is to within 63° C.

of the absolute zero. Below this there is only one thermometric substance, helium, available, so that there is no check upon the temperatures it records. But there is every reason to believe that under proper conditions the helium thermometer still registers correctly to within 3° C. of the absolute zero, the lowest temperature yet attained and measured by its aid.

There is, however, another idea involved in temperature, which, though not implied in the definition of the Absolute Scale, is implied in the quantitative aspect of the Second Law of Thermodynamics. The change of temperature of a substance is related to the quantity of heat energy it absorbs. In the simplest conceivable case, if, for a given quantity of substance, one unit of heat energy raised its temperature 1°, two units should raise it 2°, and so on. This involves the supposition that all the heat energy is employed to raise the temperature, and that the change of temperature is the only change produced by the heat in the substance. Before we can enter this gate we must previously have obtained a clear conception of what we mean by temperature and the change of temperature caused by adding or withdrawing heat energy. This we already have, so that all we need now is to specify as the ideal substance a substance of the kind described. For such a substance all the heat energy absorbed goes directly to increase the kinetic energy of molecular translation, and none is used in any other way. Now it is a most remarkable fact that the common gases, such as hydrogen and nitrogen, which are so very nearly ideal thermometric gases in the former sense, nowhere approximate the ideal substance in the sense just defined. The quantity of heat energy necessary to raise the temperature of molecular proportions of such ideal substances 1° (which would be termed the molecular heat of the substance) must be the same for all. It is very easy to calculate this molecular heat from the kinetic theory. It is three calories. Every other substance, in which some of the heat energy is not employed in raising the temperature, requires more than three calories. The molecular heat of oxygen, nitrogen, and hydrogen at constant volume (where no heat is used to do work expanding the gas against the pressure of the atmosphere) is, as a matter of fact, nearly 5.0 calories. So that for these gases only about $\frac{3}{5}$ths of the heat supplied goes to raise the temperature of the gas in the sense defined, the remaining $\frac{2}{5}$ths being spent in effecting changes on the gases other than increasing the kinetic energy of their molecular translations. What becomes of this heat? The gases considered are all alike in possessing molecules built up out of two atoms. If gases with more complex molecules are examined, like carbon dioxide with 3 atoms, ammonia with 4 atoms, and so on, it is found that the molecular heat steadily increases. As more and more atoms are contained in the single molecule, less and less of the heat goes to raise the temperature and more and more is used up in these other changes, until for complex gases, like benzene vapour containing 12 atoms to the molecule, the heat used to raise the temperature is only between $\frac{1}{8}$th and $\frac{1}{9}$th of that not so used.

On the other hand, gases containing only one atom to the molecule, the monatomic gases, are ideal substances in the sense just used. Their molecular heat is always three calories exactly as calculated. Before 1905 only one monatomic gas was known, namely, the vapour of mercury. But the inert gases present in the atmosphere discovered in 1905, of which argon and helium are the best known, are all ideal substances with a molecular heat of three calories as nearly as can be ascertained. Helium, as already noted, shares with hydrogen the distinction of being an almost ideal substance for gas thermometers. Yet with helium all the heat energy, whilst with hydrogen only about $\frac{3}{5}$ths, goes to increase the kinetic energy of molecular translation.

Monatomic molecules alone retain all the heat they receive as "sensible" heat, *i.e.* heat sensible to the thermometer. The greater the number of atoms in the molecule of a gas (which may be very nearly a perfect gas in the usual mathematical sense) the greater the proportion of heat used up, not in increasing the velocity of the molecule as a unit, but in other ways. A single atom of matter if it moves at all moves as a whole, and its energy is kinetic energy of translation. A complex structure of atoms, in addition to moving as a whole, may possess internal motion of the individual atoms with reference to one another. A colloquial way of describing these various kinds of heat energy is by the terms path-heat, spin-heat, and wobble-heat. In the first the molecule moves as a whole, in the second it spins or rotates as a whole, and in the third its parts or atoms move with reference to one another.

Now when a complex gas is heated and cooled, the heat it absorbed in the first process is given up again in the second. Therefore all these various kinds of motion of the molecule cannot be independent. There must be continual readjustments, as the temperature is rising and falling, between the path-energy, spin-energy, and wobble-energy. In general, for any mean kinetic energy of translation the molecule must possess a definite kinetic energy not of translation. In this difficult field the mathematicians have not been afraid to apply their analysis, and the law of the equipartition of energy has been extended to include equipartition, not only as between molecule and molecule, but as between the various kinds of motion possible, or degrees of freedom as the technical term is, for the same molecule. The similarity of helium and hydrogen in their thermometric behaviour, which at first sight might not appear very remarkable, in reality depends upon some of the most profound and far-reaching generalisations of mathematical physics.

Recapitulating, heat energy communicated to a material substance increases the kinetic energy of translation of the molecules as a whole, and it is only this portion of the heat energy which influences the temperature. Temperature, therefore, is a measure of one kind only of heat energy. In addition, heat produces, in all substances except the monatomic gases, many other kinds of motion, to an extent greatly dependent upon the nature of the particular substance. These motions increase with rise of temperature and decrease with fall of temperature, without producing any permanent material change of the substance, but they are altogether distinct from changes of temperature. They are connected with temperature changes indirectly, because of the universality of the law of equipartition of energy, which holds not only as between all freely moving particles whatever their mass, colliding among themselves, but also as between the various kinds of motion in space which these particles can execute. The quantitative aspect of the Second Law shows that none of these other kinds of heat energy ever can be converted into other forms without first coming into evidence as sensible heat, or the kinetic energy of molecular translation.

Obviously there is a limit to the amount of heat a complex molecule can absorb without being chemically decomposed into more simple molecules. A fly-wheel or grindstone driven at too high a speed bursts into pieces. A complex molecule similarly is dissociated.

It is customary to distinguish between two kinds of material changes, such as are produced by heat. In the first class, represented by the vaporisation of a liquid or melting of a solid, the matter returns completely to its initial state when cooled, and these were called physical changes. Whilst in the other class, chemical reactions are brought about, new molecules are produced, and these may or may not revert to their pristine condition on cooling. But although at the extremes the distinction between the two classes of change is well marked there is no definite boundary line. If the heating of water is continued, after it has all passed into steam, up to high temperature, part of the water is decomposed into its constituent elements, hydrogen and oxygen. These two gases may be readily separated at high temperature and cooled. Kinetic energy of heat has been absorbed and converted into a new kind of potential energy, the energy associated with chemical change. Hydrogen and oxygen molecules, separate from one another, but intermingling as one gas, tractate into water when the temperature is raised. Separated, they possess positional energy which, on their intimate union into the compound water, passes into kinetic energy, usually with an explosion and the evolution of much heat. The case differs in degree only from the condensation of steam into water. Very few of the best known and simplest molecules withstand dissociation into their elements when heated to 1800° C., the temperature at which the metal filament of an incandescent lamp works. Just as no explanation is possible of gravitation, none can be advanced of these cases of physical and chemical tractation, unless we ascribe "affinity" and "incompatibility" to molecules and atoms as if they were human. All that can be stated is that it is the nature of hydrogen and oxygen separated to tractate and of explosive chemical compounds to pellate. The tendency of the time is to refer these actions ultimately to the interplay between electricity and matter, but nothing very definite has yet transpired. Electricity as a distinct physical entity has been recognised scarcely more than a decade. Many of the commonest chemical changes are conditioned by the transport of electricity. It is possible that all cases of potential energy involve the action of electricity as well as of matter.

The relations between energy and matter when the matter undergoes chemical change call for some attention. After the extinction of the first glimmer of light, with the downfall of the old Phlogiston theory and the rise of the Lavoisier school, chemistry became for a time an almost purely materialistle science. It weighed everything that gravitated, analysed everything that was found or could be made, established the laws of chemical change, and founded the architectural side of chemistry—the part, that is, which deals with the precise manner in which the individual atoms are arranged within the molecule. But of energy and its importance in chemistry it understood little until the middle of last century. Nowadays, alongside of the material changes the changes of energy are studied as a matter of course, and new applications of the laws of thermodynamics are being made, with results as valuable and as unsuspected in many cases as those which followed their original application to simple changes of state such as fusion and vaporisation. A few examples may suffice of some of the important developments of recent years. In their search for cheap energy, chemists have invaded even the most out-of-the-way places, and nothing is more common, among districts visited a few years ago only by tourists in pursuit of the scenery of lakes, mountains, and waterfalls, to find small chemical factories dotted about, nourished by the "white coal," as it is termed abroad, the water power before running to picturesque waste. Around Niagara Falls large towns have sprung up connected with chemical industries requiring energy, whilst the valleys of Switzerland have been described as "glacier at one end and 98% nitric acid at the other."

There are two kinds of processes in chemistry which are with advantage carried out in conjunction with the large modern power stations. The first are those requiring exceptionally high temperature, and the second are those requiring cheap power. For the power, whatever its source,—whether waterfalls, natural gas or oil, waste gas from blast furnaces, or cheap coal at the pit's mouth,—is always first converted into electric energy by means either of large turbines in the case of water, or of gas-engines in the other cases. As electrical energy, it is transmissible at high voltages cheaply over a considerable area, and is readily and efficiently transformed into the particular kind of electric energy most suited to the process served, or into mechanical energy.

The first class of process, requiring exceptionally high temperature, is carried out in the electric furnace, where the current passes either as an arc within the furnace, or through a high resistance, generating any temperature up to from 3000° to 3500° C. A whole host of entirely new compounds has thus been prepared, and many have important applications. Calcium carbide, which gives acetylene on addition of water, is a compound of one atom of calcium and two of carbon, obtained by so heating lime and coke. Silicon carbide, or carborundum, from sand and coke is used largely as an abrasive and polishing material in place of emery. Moissan made his microscopic diamonds in the electric furnace, and numerous and unsuccessful have been the attempts of others to improve upon his results. If diamonds were common instead of costly, what magnificent tools could be made for certain purposes and how certain industries like civil and mining engineering, engaged in drilling through the solid rock, would blossom forth! This is no idle dream like the making of gold, and the age to succecd the steel age in the future may yet be a diamond age. The diamond as a gem is beautiful, but as a tool it is a power, on account of its magnificent hardness and resistance to wear.

The second class of industries, which is concerned with substances requiring a large amount of energy for their production, although not necessarily distinct from the first class, have often different objects. The first class produce substances either previously unknown or very rare, which often cannot be produced in any other way. The second produce well-known commodities often in universal demand by new processes in which the energy is more directly introduced than in the old. For certain reasons the element nitrogen figures most largely in this class. The element nitrogen is one of the most interesting to chemists and one of the most important in the processes of life. It has at first sight a paradoxical nature. Except for the monatomic gases of the atmosphere, it is the most inert element known, entering into combination with other elements very reluctantly, and yet forming a very large number of compounds, such as ammonia, nitric acid, and their salts, which are in universal demand. Free nitrogen from the atmosphere is to be had for nothing in unlimited amount, but combined nitrogen, as in the ammonium salts and nitrates, commands a relatively high price, and the demand for it will increase whilst the present sources of supply must ultimately become exhausted. Nitrogen is the essential element in almost all explosive compounds. It forms a numerous class of unstable derivates, which on a slight disturbance detonate, the element nitrogen being liberated in the free state as a gas with sudden enormous expansion of volume and the accompanying destructive effects. Nitrogen is peculiar, probably because its atoms naturally tend to combine with one another, rather than with other elements, to form the very stable diatomic molecule N_2. Hydrogen and oxygen also form similar molecules, H_2 and O_2, but in a mixture of these gases, a spark or any local rise of temperature is sufficient to bring about a rearrangement of partners, the atoms before combined with atoms of the same kind then uniting with atoms of the other kind. Single atoms of nitrogen never exist uncombined, even at high temperature, on account of their tendency to unite with one another, and so are not known. But there is good reason to believe that as much energy is liberated when free single atoms of nitrogen unite to form the diatomic molecule of nitrogen as in any other chemical change known. Hence, before the worthless free nitrogen of the atmosphere can be converted into the valuable combined nitrogen, so essential for agriculture and the production of all food-stuffs, this energy must be supplied. Free atoms of nitrogen tractate with an unparalleled conversion of potential into kinetic energy, and before the molecule of nitrogen, consisting of two atoms united together, can be made to form other compounds, such as ammonia and nitric acid, the two atoms must be separated again. In this process the same amount of kinetic energy disappears and is reconverted into energy of position. The modern industrial development for the utilisation of atmospheric nitrogen dates from but a few years back. In principle no essential advance has been made upon Cavendish's great discovery in the eighteenth century that, when an electric discharge is pressed through air, a small part of the oxygen and nitrogen combine to form nitrous acid. Instead, however, of the energy of a patient attendant turning the handle of a frictional electrical machine, hour after hour, as in Cavendish's experiment, to-day the power of hundreds of thousands of horses, derived chiefly from the "white fuel" of the Norwegian and Swiss hill-sides, are ceaselessly at work, turning dynamos which produce powerful high-tension arcs in the air, so converting it partially into nitrous and nitric acids. Much of the product is neutralised with lime, and the calcium nitrate resulting is employed as a fertiliser of the soil in agriculture. It is an interesting example of a chemical industry in which the raw materials cost nothing or next to nothing, and all that has to be supplied to convert them into products of value is energy. The conversion takes place in Nature's own slow way in the air and soil, but modern civilisation could never subsist on this parsimonious allowance. Energy it must have just in proportion as it advances, and the day-to-day supply of sunlight which sufficed for the requirements of primitive agriculture is insufficient for modern intensive methods.

It is interesting to trace the source of the supply of combined nitrogen on which old-time agriculture depended. First, a very slow oxidation of the nitrogen in the air takes place whereby about 11 lb. of combined nitrogen are formed per year per acre of ground and carried down by rain. It used to be thought that this process was effected by lightning and other discharges of electricity in the atmosphere, but this was before the days of radium. The atmosphere contains the emanation of radium, which, though in relatively infinitesimal amount, aggregates in the total all over the globe to a very considerable quantity, corresponding with hundreds of *tons* of pure radium. The rays from this radium emanation are very effective in causing oxygen and nitrogen to combine, and no doubt the supply of combined nitrogen from the air is largely due to its action. Secondly, certain leguminous plants, by the aid of bacteria in nodules in the roots, have the power of converting atmospheric nitrogen into the combined form, and the empirical principle of the rotation of crops depends on this action. But other bacteria possess exactly the opposite power, and undo the work of the nitrifying organisms. It is doubtful if this process accounts very largely for the maintenance of combined nitrogen in the soil.

This industry has another interest in that it is the first in which any uneasiness has manifested itself as regards the future. Some years ago Sir William Crookes, in a Presidential Address to the British Association, drew attention to the probable future failure of the wheat supply. In addition to manure, the only important sources of combined nitrogen available, now that the guano deposits are exhausted, are the natural deposits of sodium nitrate or Chili saltpetre ("caliche") from the rainless districts of S. Peru and Bolivia and the ammonium salts obtained as by-products in the manufacture of coal-gas. The latter is largely wasted owing to the extravagant and uneconomical methods in vogue of utilising coal and could be enormously improved. Both these natural resources, Chili saltpetre and coal, are far from inexhaustible. The former, it is estimated, will only suffice from twenty to forty years longer at the present rate of consumption. The boisterously prosperons last half-century, in which virgin territory was being everywhere opened up and the conservation of natural resources was hardly thought of, is giving place to a period of reflection in which awkward interviews between civilisation and its banker are in prospect. The first suspicion that a day of reckoning was to follow arose out of the address of Sir William Crookes, which pointed out that the wheat supply would fail unless chemists succeeded in solving the problem of the fixation of atmospheric nitrogen. Note the result. Great corporations have spent millions of pounds in the processes for utilising atmospheric nitrogen. The problem has been practically solved, and no doubt when the Chili saltpetre supplies fail, or earlier, they will reap a rich harvest. But how is it done? Simply by the proper expenditure of natural energy, so that the question, as to how long the Chili saltpetre beds will last, has simply been merged into the more general problem of how long the natural resources of energy of the globe will hold out. In so far as such developments utilise the natural energy running to waste, as in water power, they may be accounted as pure gain. But in so far as they consume the fuel resources of the globe they are very different. The one is like spending the interest on a legacy, and the other is like spending the legacy itself. The wheat problem, with its uncomfortable suggestion of impending starvation for a large part of the race, is one particular aspect of a still hardly recognised coming energy problem, which to thinking minds perplexes the whole future and is like "the cloud no bigger than a man's hand" on an otherwise cloudless horizon.

One point remains before leaving the study of heat and chemical energy. Why is heat energy of uniform temperature the ultimate fate of all energy? The power of sunlight and coal, electric power, water power, winds and tides do the work of the world, and in the end all unite to hasten the merry molcecular dance. Those who have followed the line of thought that has been pursued will have little difficulty in answering this question. There is a definite stop to the progress of subdivision of matter, when the single atom or the single molecule, according as chemical or physical methods of division are employed, is attained. There is a definite limit to the process of the transformation of energy, when the subdivision and distribution of that energy, both in quantity and direction, among the smallest possible portions of matter has been achieved. The limit which puts a stop to the one process necessarily puts a stop to the other. If molecules were merely very small particles of matter, rather than the smallest possible single units, and if still smaller particles existed, the collision of two molecules, which for every molecule in a gas under ordinary conditions takes place hundreds or thousands of millions of times every second, would result in loss of kinetic energy of motion by reason of the imperfect elasticity of the matter and the subdivision of the part of the energy lost among these still smaller particles. That is to say, if molecules were not a real limit of subdivision, molecular motion would not be the last stage in the transformation of energy.

The definition that the molecules and atoms are the ultimate limit of the subdivision of matter by physical and chemical agencies carries with it the necessary corollary that molecules and atoms must be perfectly elastic and frictionless. Friction and imperfect elasticity are properties of the gross world, and express the transformation of orderly motion into commotion. With the molecule and the atom we reach the limits of subdivision of matter. There are no smaller material particles among which the energy can be subdivided, and therefore we reach the limit of subdivision of motion.

By following these lines of inquiry a clear and definite mental impression of the nature of heat has been gained. How external a phenomenon it is to the molecular world. Thermal and mechanical energy are one and the same thing to the individual molecule. It is true that at sufficiently high temperature heat will bring about the decomposition or dissociation of most complex molecules, and probably, if the process were pushed far enough, would resolve most common gases into single atoms. But when we have reached this final limit of subdivision we have no reason to believe that at any temperature, or under any circumstances, any of the heat energy is transformed from motion of the atom as a whole into internal commotion. Were it otherwise, at a sufficiently high temperature the atoms themselves would be dissociated, and transmutation would take place under the action of heat. This was the alchemist's dream. Furnaces and crucibles were his main stock in trade. Nowadays by the use of the electric furnace we can attain in a few minutes temperatures almost incomparably greater than the utmost the alchemist, after days and nights of anxious care, could attain. But transmutation does not seem to lie that way. Inside the atom, heat energy does not exist, and temperature has no meaning whatever. Under transcendental conditions —for instance, in the sun and stars—dissociation of the elements into simpler forms has been imagined to be, and may be taking place, for, at these dazzling temperatures, radiation, a phenomenon which in our study of low temperature heat we have not yet had occasion to consider, is of paramount importance, and radiant energy is very different from the energy of motion of molecules. However, the fact remains, that so far, even at the highest attainable temperature rendered available by the use of the electric furnace, no indications of a transmutation of the elements is yet forthcoming.

Yet the atom, for all that, is not Nature's unit, but ours. Of recent years radioactivity has been traced to the spontaneous disintegration of certain atoms into parts which form lighter atoms. The energy evolved in this natural change puts into the shade every previously known example. Mass for mass, the most violent explosives known, in which suddenly the atoms composing the molecule pellate, liberate scarcely a millionth part of the energy set free when atoms fly to pieces. The great difference between the two cases—between, that is, the most fundamental kind of material change known ten years ago and that known to-day—has opened up for science an entirely new horizon. Kinetic energy only is sensible and knowable. Potential energy may and does exist in matter to an extent even now scarcely capable of being grasped, but until the matter changes and its energy of position is converted into energy of motion, this energy is unknowable and unavailable.

CHAPTER VI

ELECTRONS AND X-RAYS

To the early philosophers electricity was a "fluid," weightless and immaterial, it is true. Concerning its real nature there was preserved a discreet silence, so that it has been remarked that any too great curiosity about it almost came to be considered indelicate. But it was a kind of fluid, in that it had the power of flowing through metals and other conductors of electricity, much as water flows through a pipe. Even the best conductors of electricity offer resistance to the flow of electricity, just as the smoothest and widest pipes offer resistance to the flow of water, and some of the electrical energy of the moving electricity in the one case, just as some of the mechanical energy of the flowing water in the other, is, during the passage, converted into heat. If the wire through which the current flows is of small cross section and is made of a substance which is not a particularly good conductor, such as carbon, then the heat produced by the friction to the passage of the electricity may heat it to whiteness and cause it to emit light, as in the filament of any electric lamp. If the mains actually touch without any resistance to the passage of the current, a "short circuit" results. If not protected the mains would instantly melt. So to prevent this "fuses" or thin wires of lead are introduced at suitable points, which melt if the current increases above a prearranged value, so protecting the mains from injury. All the applications of electricity to incandescent lighting, heating, and cooking depend on the principle just discussed. There is a generating station from which the electricity is pumped out into one main by a dynamo. After passing through the lamps or heating apparatus it flows back to the station through a return main, just as water might be pumped round a system of pipes. The duty of the staff of the generating station is simply to maintain the pressure or head of electricity—technically called the voltage of the supply—constant whatever quantity of electricity is flowing round. If the demand increases more machines are started and put into duty, for whatever the demand the pressure at which the currents is supplied must not vary. By making the mains of copper of the highest possible conductivity and of sufficient cross sectional area, so that in them the resistance losses are small, and the filaments of the lamps or wires of the heating apparatus of just the right length and cross section to let the amount of electricity pass at the prevailing pressure or voltage to do the work required, by far the greatest part of the electric energy supplied is converted into heat or light at the place wanted, and only a small part is dissipated *en route*. The Board of Trade unit at which electric energy is retailed to consumers in this country is the "kilowatt-hour"—that is, a thousand "watts" acting for one hour. The unit of pressure is the volt, and the unit of current is the ampere. The product of the volts and the amperes expresses the power in watts, and the product of this again into the time the current flows the amount of electric energy consumed. The heat equivalent of the watt is 0.24 calories. One Board of Trade unit of electrical energy if converted into heat will raise 15.2 pints, or nearly 2 gallons of water from the freezing to the boiling-point. The carbon filament incandescent lamp requires about $3\frac{1}{2}$ watts, whereas the new metal filament lamps require only about 1 watt, per candle-power of light given.

In this country, electricity is almost always generated from coal, and the chemical energy of this is first converted into heat of only moderately high temperature, generally in steam boilers, before it is converted into electrical energy. Since this must be, by the laws of thermodynamics, a wasteful process in which the greater part of the whole heat energy goes down the drain, in the form of low temperature heat with the condensor water from the steam engine, it follows that only a small part of the original energy of the coal finds its way to consumers' houses as electrical energy. The conversion of this into heat by use is complete, a given consumption of electric energy giving in the end a definite amount of heat, however it is employed. For this reason electric heating must always be costly as compared with direct heating by coal. In electric lighting the amount of energy converted first into light is very small compared with that converted directly into heat, and considerable improvement in the efficiency is possible in this direction. The part converted into light, or radiation, increases very rapidly as the temperature of the filament of the lamp is raised, the light at the same time becoming more nearly white. The modern metal filament lamps, made out of the rare metals osmium, tantalum, and tungsten, run at a temperature some hundreds of degrees hotter than the older carbon filament, and their increased efficiency and the whiter character of their light are due to this higher temperature. But in electric heating and cooking no improvements in the efficiency are possible. All that can be done is to generate the heat just where it is wanted and to prevent it from getting out and being wasted as much as possible.

The simple idea of an immaterial fluid capable of being pumped under pressure through a metal wire, like water through a pipe, suffices, without any deeper analysis, to explain in a way some of the most everyday applications of the electric current. There is, however, one point which is not covered. We can tell which way the material water is flowing in a stream by mere inspection, but we cannot tell which way the immaterial electricity is flowing. It is true we label one main as the positive and the other as the negative for the sake of distinguishing them, and there are certain easily distinguished differences between the + and − wires. For example, when immersed in slightly acidulated water, hydrogen is evolved at the negative and oxygen at the positive wire by decomposition of the water; and since the volume of the hydrogen is twice that of the oxygen, and since unless the wire is of a "noble" metal, like platinum, most of the oxygen at first liberated combines with the metal forming an oxide, hardly any escaping as gas at all, no other indication is necessary to distinguish between the + and −. But does what we call electricity flow from the + to the − or *vice versa?* We need to go a good deal deeper to answer this question.

Two rival views held the field with the older philosophers, whose ideas originated in observing, not currents of electricity, but stationary electric charges. Some substances, like glass rubbed with silk rubbers, became electrified with one kind, positive as they conventionally termed it; others, like vulcanite rubbed with flannel or cloth, became electrified with the opposite kind of electricity, which was termed negative. Similar charges were stated to "repel" one another, whereas opposite charges "attract" one another. Some supposed that there was but one electric fluid, which was a universal constituent of all matter, whether electrified or not, the state of positive electrification being when the matter possessed either less or more than the normal amount, the state of negative electrification being the reverse. Others supposed that two kinds of electricity existed, which were the exact opposite counterparts of one another, something analogous, possibly, to an object and its mirror image. All electrical phenomena can be explained as well on the one fluid as on the two fluid idea, but our ignorance at the present time as to whether there are two kinds of electricity or one is fundamental. Until the question is settled, the hopes that have been entertained that, through the study of electricity, we shall be able to arrive at a philosophical explanation of matter, are likely to prove unfounded.

Now, what is the revolution of ideas which the closing decade of last century witnessed? The material fluid, water, to the eye of the poet, the symbol of peace and rest, its flow a quiet, continuous, gliding movement, viewed through the molecular spectacles of science presents a picture, compared with which the most frenzied struggles of a fighting mob is almost absolute stillness. So the electric fluid, when it is forced into the limelight of searching inquiry, undergoes a similar transformation. No more than matter does it occupy space continuously. It consists of a myriad of separate small units or *atoms* each leading an independent existence, like individuals fighting for their own hands. Whereas, however, we know over eighty different kinds of material atoms, we know only one atom of electricity; and this is the kind of electricity which is, by convention, termed negative electricity. If there is only one kind of electricity, and only one so far has been isolated from matter, it is of the kind which, unfortunately, came to be termed negative. The atom of negative electricity, which like the atoms of any one element are all precisely of the same kind, and so far as is known are not divisible into smaller units, is termed the *electron*. The electron is the smallest entity known to science. Just as the material fluids, liquids and gases, are made up of a vast swarm of minute particles, so the electric fluid, the flow of which in a conductor produces the electric current, is made up of a vast swarm of separate electrons. The current through a wire may be likened to the passage of these negative electrons, which are known, from the negative pole to the positive pole, or in the opposite direction to that conventionally assumed. If positive electrons also exist, and take part in the flow of the current, these would, of course, move in the opposite direction. But so far the positive electron, though much sought after, has eluded pursuit; and however necessary it may be for the electrical theory of the nature of matter, for all purely electric and electromagnetic phenomena, as already remarked, one kind of electron explains the known facts as well as, or possibly better than, two. For mere purposes of description it is briefer to assume the existence of both kinds.

The electron was first recognized as a separate entity only after it was isolated from matter. Just as a charged piece of sealing wax "attracts" a piece of paper or other light object, so a free electron by virtue of its being an electric charge "attracts" a molecule and attaches itself to it when it gets the opportunity, forming a negatively charged molecule or *ion*. The positive ion is always molecular in size, and is usually regarded as an electrically neutral molecule which has lost one or more of its normal number of electrons. But the negative ion, if produced in the absence of all molecules, in an almost perfect vacuum, for example, is not of molecular size, but an entity 2000 times lighter than the smallest known atom, the atom of hydrogen.

The term *ion* was introduced by Faraday to explain the facts of electrolysis, or the chemical phenomena which occur when the electric current is passed through a solution, usually in water, of a compound substance. The current was regarded as being carried through the liquid by the movement of molecules charged with electricity. The positively charged ions moved to the negative pole and there gave up their charges, the molecules being deposited, whilst the negatively charged ions moved to the positive pole and likewise gave up their charges, their molecules being similarly set free. The flow of the negative charges in the one and of the positive charges in the other direction constitutes the electric current. Water itself is not a good conductor, but solutions of common metallic compounds usually conduct, the dissolved substance becoming decomposed, the metal being deposited in the form of an adhering film on the negative electrode, or pole, by which the current is conveyed to the liquid. This is the basis of all the arts of electroplating with gold, silver, nickel, copper, etc. Now

Faraday found that, comparing two metals like copper and zinc, the atomic weights of which are 63.6 and 65.3, the current that would deposit 65.3 grams of zinc in one solution, say of zinc sulphate, would deposit 63.6 grams of copper in another solution, say, of copper sulphate; whereas for silver, the atomic weight of which is 108, the same current led through a silver sulphate solution would deposit, not 108 grams of silver, but exactly twice this amount. Now silver is a monovalent metal and can combine with only one atom of another monovalent element like chlorine, whereas copper and zinc are, in the compounds stated, divalent and will combine with two atoms of chlorine. Another series of copper compounds is known, however, in which the copper is monovalent, like silver, and, if these are electrolysed, twice as much copper is deposited from them as by the same current flowing through solutions of the commoner series of copper salts. On the atomic theory of electricity how simple these phenomena appear, and yet how long was it before the explanation was applied! Faraday's ions are compounds of the atom or groups of atoms with an atom of negative or positive electricity, or an electron less or more than the normal number, in the case of the monovalent ions, with two electrons in the case of the divalent ions, with three in the case of the trivalent ions, and so on. To liberate an atomic weight in grams of a metal requires that just as many electrons should be sent through the solution, as electric current, as there are atoms in this quantity in the case of a monovalent metal. For divalent metals twice as many electrons are necessary, and so on, each material atom transporting one, two or more electrons, but never fractions of the electron. But until the electron was isolated from matter nobody had the courage to believe in it as a separate entity. New experimental methods had first to be perfected for obtaining space free from matter, or in other words a perfect or almost perfect vacuum.

Toricelli's famous experiment of filling a glass tube completely with mercury and inverting it in a bowl of mercury, whereupon the mercury fell in the tube to the barometric height, about 30 inches at sea level,—leaving a perfect or almost perfect vacuous space at the top of the tube, has been adopted in the modern mercury pump as an almost perfect means of removing the air from a closed vessel. When the electric current is forced through the gas in such a vessel, connected with the air pump, while the air is being gradually removed, an extraordinary sequence of beautiful phenomena present themselves. At first, before the exhaustion commences, it requires an electrical pressure of tens or hundreds of thousands of volts to cause the current to flow at all, and then it does so like lightning, as a disruptive spark making a sharp noise. But as the air is pumped out the current passes with much greater facility, and requires only a few thousand volts. The spark gives place to the glow of the vacuum tube, and the discharge broadens out until it gradually fills the whole vessel completely. At this stage the current is carried by positive and negative ions, or charged molecules of the gas, streaming in opposite directions to the two electrodes, their collisions with one another probably causing the light. A closer examination at this stage will reveal the fact that the main resistance to the passage of the current is at the boundary of the negative electrode and the gas. That is to say, the electrons flowing in the metallic part of the circuit from negative to positive, leave the metal and jump off into the gas only with great reluctance, the resistance everywhere else in the circuit to the passage of the current being relatively small. A mercury surface in a good vacuum, although it opposes this jumping off of the electrons as much as the other metals do at first, has the peculiarity that at the moment the current starts the reluctance practically disappears. Once started, the current, even although at quite low voltage, continues to flow through the mercury vapour with ease. This is the principle of the mercury lamp, which will run on the ordinary electric mains. It is started by tilting the lamp, so that for a brief moment the mercury flows from one pole across to the other, causing a discharge to pass; and, when once so started, the current will continue through the mercury vapour, producing the well-known ghastly coloured but cheap and otherwise pleasant light. Other means of overcoming this great reluctance of the electrons to leave the metal and to enter the vacuum (which is known technically as the cathode fall of potential) are known. For example, the effect may be achieved by covering the negative pole with lime and making it white hot (the "Wehnelt cathode"). It is not difficult to predict that probably the electric lamp of the future will be some form of vacuum tube capable of being run from the ordinary electric mains. The beauty and brilliance of the vacuum tube discharge through gases when powerful currents from the mains at low voltage are used, instead of the very small currents at high voltage furnished by the induction coil, can only be appreciated after having been seen.

If we continue our experiment we shall find that when all but from about 1/1000th to 1/10,000th part of the air has been removed, the current flows with the minimum of opposition, and from this point, as we go on exhausting, the resistance to the passage of the current again increases and goes on increasing very rapidly until, finally, the most powerful appliances known quite fail to force the discharge through. At this stage the smaller the distance between the electrodes the more difficult it is to force the current through, and, at a degree of vacuum easily reached nowadays, the discharge may prefer to leap across several inches of the air outside the vessel rather than pass through an inch distance of the vacuum within.

This is the stage of vacuum from the study of which so much that is new and revolutionary has been gleaned. It can only be attained by a mercury pump or equally efficient modern methods, and this accounts for the fact that the older observers, with mechanical air pumps, were never able quite to obtain the most interesting stage of the vacuum. The discharge alters its character, and interest centres around the negative electrode, or "cathode," as it is termed. The positive electrode or "anode" may have any position or shape without any longer affecting the character or direction of the discharge. From the cathode and at right angles to its surface, there issue now the electrons or atoms of negative electricity in the free state, unhampered by molecules of matter and travelling alone. Sometimes, if the vacuum has not been pushed to its extreme limit, the paths of these electrons ("cathode-rays" or "cathode-streams," as they were first and are still called) can be faintly seen, but usually it is not until they impinge upon an obstacle that they become evident. If a small piece of platinum foil is put into their path, it can be heated white hot and may readily be melted. The glass walls of the vessel, wherever they are bombarded by these streams, fluoresce, usually with a beautiful pale greenish glow; and if an object of definite geometrical form, like a cross, is arranged in the vessel so as to intercept part of the stream, it casts a shadow of its own form, the glass behind not fluorescing where it is protected from the bombardment. The electrons travel outward from the cathode in straight lines, everywhere, as a first approximation, normal to the surface of the cathode, and so they resemble "rays," and were, in fact, named "cathode-rays" before their nature was elucidated. Certain fluorescent substances, particularly a silicate of zinc known as willemite, and the class of salts known as the platinocyanides, for example, platinocyanide of barium, respond to the cathode-rays far more brilliantly than glass; and these substances when put within the vacuum tube produce the most beautiful effects. Sir William Crookes examined these new phenomena exhaustively, and, without actually elucidating their true nature, he came near enough to assert that the effects were produced by swarms of material particles in a new or fourth state, not solid, liquid, or gaseous, but something ultra-gaseous, or radiant. His name "Radiant Matter" still survives along with the terms cathode-rays and cathode-streams.

No subject was a more favourite one for lecture experiments and investigation, but, curiously, all who worked with these highly exhausted vacuum tubes until 1905 overlooked probably what would be considered from a utilitarian point of view the most extraordinary phenomenon of all. When these electrons are forced out of the negative electrode, they rush away from it, by the law of the pellation of similar electric charges, along straight lines like rays. When they strike a solid obstacle, they are converted into X-rays, or Röntgen rays. The denser the obstacle, and therefore the more suddenly the electrons are stopped, the more penetrating and remarkable are the X-rays produced. The X-ray tube is simply a Crookes' tube, with an obstacle of the dense and difficultly fusible metal platinum, called the anti-cathode. The cathode is made concave, whereby the electrons it emits are brought to a focus in the middle of the tube, at which point the anti-cathode is placed. Of the application of the X-rays, discovered by Röntgen, and of their power of penetrating matter opaque to light, it is scarcely necessary to speak, nor of their invaluable aid in surgery and diagnosis. They enable the physician actually to see the denser parts of the interior of the body, and to locate fractures and certain growths. In fact, they furnish him with a veritable sixth sense.

What creatures of circumstance we are, awakening into consciousness in a complicated world, and apprehending the simple only through much less simple but more obvious phenomena. Obvious inexplicable things force entrance even into our philosophy and silence inquiry. Why is glass transparent and metal opaque? Until the X-rays were known, to which metals are not specially opaque nor glass specially transparent, the complicated peculiarities of light, which traverses some materials practically unabsorbed, and is completely stopped by the thinnest films of others, seemed most natural and commonplace. The recognition that there was nothing absurd about a particular sort of light traversing a sheet of iron or stone, which arose with the discovery of the X-rays, was extended by the discovery of the α-, β-, and γ-rays of the radioactive substances, and by the nearer study of the cathode-rays. The latter, although not sufficiently penetrating to escape from the vacuum tube, yet in their general absorption by matter, do not recognise the optical qualities of opacity and transparency, any more than do the X-rays. In the absorption of these newer types of radiation we have, as a first approximation, the simplest conceivable law. The power of penetrating matter varies enormously with the different types, from that of the α-rays, which are completely absorbed by a sheet of writing-paper, to that of the γ-rays of radium, only half absorbed by half an inch of solid lead. But, for all, the quantity or mass of matter in the path of the ray is what fixes the amount of absorption, and considerations of the physical and chemical nature of the matter hardly enter. All substances absorb these new rays proportionally to their density as a first approximation, when sheets of the same thickness are compared. The penetrating power of the X-rays varies very much with the degree of vacuum of the X-ray tube. If the vacuum is poor or "low," the rays are not very penetrating or "soft"; whilst if the vacuum is good the rays are penetrating or "hard." But the most penetrating X-rays fall far short of the γ-rays in this respect, and are never able to penetrate more than a very small fraction of an inch of lead.

CHAPTER VII

INERTIA

In attempting to follow further the foregoing line of thought, we are abruptly pulled up because, so far, we have been concerned merely to make electricity, like matter, atomic in structure, and to make the electric fluid nothing more than a particularly fine-grained sort of gas, without any reference at all to properties which the electron has, and which no material particle possesses. We cannot proceed more than a very little way in the study of electricity by the use of material analogies, any more than we could pursue the study of bacteria if our minds were able to apprehend nothing less highly organised than a man. The average mind will follow easily the advances of science so long as they may be expressed in terms of mechanical models, and by means of mental pictures of concrete things. But when we leave the realm of matter, and attempt to penetrate into that of electricity and the ether, the highest intellect feels the need of models and the impossibility of obtaining even the raw unfinished material out of which to construct them. It is as if we were in a world destitute of simple wood and brass and nails, but elaborately furnished with all sorts of extremely complicated constructions which baffle our ingenuity to pull to pieces. Our most fundamental conceptions are, like ourselves, material. The elaboration of them is easy, but their simplification to suit the immaterial world, whither we now wish to embark, is difficult almost to impossibility. If our minds habitually thought in terms of electricity and magnetism instead of in terms of matter and motion, what a world would be opened up!

We have first to attempt to realise that an electrón is an electric charge which, if not prevented, pellates from other similar electrons or negative electric charges, and tractates towards positive charges. The intensity of the effect varies, like that of gravitation, inversely as the square of the distance. If a collection of electrons is brought near an uncharged mass of matter of finite size, the electrons in the matter pellate to the further side, leaving the side nearest to the electron positively charged (by induction, as it is called). The opposite charges being nearer than the like charges, the tractation overpowers the pellation. Uncharged masses in the proximity of electric charges of either sign tend to approach these charges. These effects are very much greater than any of mechanical origin. To fix our ideas we must first have in our minds some definite quantity of electricity or some definite number of electrons. Suppose we take as many electrons (7×10^{23}) as there are atoms in a gram of hydrogen or in 108 grams of silver. This is the number conveyed through a circuit by a current capable of depositing 108 grams of silver from a silver solution in an electroplating bath. It is roughly the number flowing in four days through the filament of a 16 candle power carbon incandescent lamp burning on a 200 volt circuit, or through an ordinary 10 ampere arc-lamp in 2 hours 41 minutes. If two such quantities of electricity were placed, one at the North and the other at the South Pole of the earth, they would tend to pellate so strongly that, even at this distance, a fairly thick steel cable, capable of supporting the weight of 35 tons, would be necessary to keep them from moving apart. Since the pellation decreases as the square of the distance, if two such quantities were placed near together no material bond would hold them. The obvious conclusion since electrons have these properties, is that it would be quite impossible ever to keep the number supposed in the illustration together in one place, for every one would fly away from every other, bursting into fragments any material obstacle to their motion.

Hence, when we wish to construct an electrical model, and begin on the raw material, electricity, we are confronted at once with a set of conditions quite alien to anything with which we are familiar. The brass and wood and nails of our mechanical models, when collected together, remain where they are put, whereas free electricity, if not prevented, dissipates its constituent particles as far as possible from one another. The moral of this illustration is that it is impossible to have many free electrons in any mass of matter. When a small ball is charged negatively to the highest possible point, so that the charge is leaking away from it on all sides through the air in the form of brush discharges, the number of free electrons it possesses is not greater than one for every million million atoms. The charge we have taken as a unit, namely, that carried by one gram of hydrogen ions, would, if free, charge the whole world up to a potential of a million volts. Free electricity, uncombined with matter, is wonderfully potent, and in the most powerful manifestations of it, the actual quantity involved is very insignificant.

Although very considerable numbers of electrons must flow to produce even a feeble *current* of electricity, the greatest charge of stationary electricity known to us is in reality an exceedingly small quantity, and we can never apprehend free electricity uncombined with matter in anything but infinitesimal amounts. The key to future progress, as already remarked, is the answer to the question, "What is positive electricity?" In other words, what controlling agent is it which neutralises the pellation of the electrons in a piece of electrically neutral metal, and, but for which, the contained electrons would be explosively expelled to the ends of space, carrying the metal with it, if they could not move without it? If, in the metal, there were an equal number of positive electrons, the two sets would neutralise one another, but why then is the positive electron never obtained free? One cannot but feel that the answer to this question underlies the secret of the structure of matter, and that the expression positive electricity is but a mere term cloaking total ignorance of the essential character of the structure the electrical theory of matter attempts to explain.

The electron has been spoken of as immaterial in the sense that it is not matter, but something at once finer grained and endowed with fundamental qualities which distinguish it from any of the known kinds of matter. But electric energy has also been spoken of as the kinetic energy of moving electrons. Kinetic energy is measured by half the product of the mass into the square of the velocity. Any moving particle, whether it is material or not, possessing kinetic energy and able to agitate the molecules of matter, or, what is the same thing differently expressed, capable of being converted into heat, must possess mass. A massless particle would belong to that other world of spirits and dreams, the inhabitants of which are not "conserved," and the study of which belongs not to physical science. A massless particle, so far as can be seen, if it in any way acquired energy, however infinitesimal, would move at infinite velocity, and would, therefore, leave the universe behind it without the lapse of time. Whether or not future generations will find any room for massless particles in their philosophy, the present can hardly conceive them to exist, or imagine how they could become known if they did. It is just because the electron has a definite mass, even though it is by far the smallest of any known, and still is not a material particle, that its chief interest lies. Assuredly science has here penetrated one step farther into the eternal verities. With many of the feelings of an airman, who has left behind for the first time the solid ground beneath him, let us try to venture into this new region of science, of mass without matter.

Mass and matter are to us almost synonyms. Indeed, mass has been defined as quantity of matter. True to the fatal habit of accepting obvious things as fully explained or fundamental, few will ever have set themselves to abstract the idea of mass from matter, so that this must first be done. We must first take care not to confound mass with weight to which it is proportional at any one latitude. The law of gravitation acts universally on matter endowing it with weight due to the tractation of the matter towards the immense mass of the earth, from which we can never get far away. If we could remove to a place in space remote from worlds of all kinds, the weight of the matter would practically disappear, but its mass would be unchanged. Under such conditions we should bring into prominence the true fundamental attribute of matter, its inertia, or its disinclination to move when at rest, and its disinclination to stop moving after it has been started. Moving it possesses energy, the kinetic energy of motion. Before it can start energy must be supplied to the matter, and before it can stop this energy must be taken from it. Unless it loses its energy it will continue to move, if moving originally, and its motion will be uniform and perpetual. It seems the most natural thing in the world that matter should possess this inertia, so natural indeed that to ask why might seem a foolish question. At least the question was never asked before the electron was studied and it was found that the inertia of an electron, the property by virtue of which a moving electron possesses kinetic energy, accounted also for a great deal more than its inertia.

Electric currents, which we have seen are due to the flow of electrons, are not merely moving charges of electricity. So far, attention has been confined practically to the heating effects of the current, or to the communication of the motion of the electrons by friction or resistance to the molecules of matter. A charge of electricity, or electrons at rest, has no *magnetic* properties. A current of electricity, or the same electrons in motion, has. The electrons flying free from matter in the Crookes' tube are turned out of their course when one pole of a bar magnet is brought near to the tube. Faraday imagined that "lines of magnetic force" extend out from a magnet into space. These lines, which we shall simply refer to as the magnetic lines, merely point the direction along which magnetised bodies tend to approach a magnet. In the case of a bar magnet the lines continue out in the direction of the length of the magnet. If a narrow pencil of cathode-rays is examined, it will be found that when the bar is held so that its length is at right angles to the path of the rays, the latter, when they pass the end of the pole, are turned to one side in a plane at right angles to the length of the bar magnet. If the other pole of the magnet is then presented, the rays will be turned just as before, but to the other side. But these same electrons at rest, as a simple electric charge, are not moved, nor affected in any way whatever by a stationary magnet. It is true that they are affected by a moving magnet, but a magnet moving with reference to an electron comes to the same thing as the electron moving with reference to a stationary magnet. If the two poles of a magnet are bent round so that they are opposite each other, the magnetic lines are in the direction of the shortest straight lines joining the two poles. A cathode-ray travelling up between the poles at right angles to the magnetic lines is, if the magnet is strong enough, coiled into a circle having one of the magnetic lines as its axis. If it is not travelling at right angles to the magnetic line, it pursues a spiral path around the magnetic line, its component of velocity along the magnetic line not being affected. If it is travelling in the direction of the magnetic lines, its motion is unaffected by the magnet. In the first case referred to, the electrons assume a new motion, the direction of which is at right angles to the magnetic lines and at right angles to the original path of motion of the rays. We live in a world of three dimensions—length, breadth, and depth. If the direction of the electron's original motion is along the length, and that of the magnetic lines along the breadth, the electron departs from its original path with a new motion in the direction of the depth. The diameter of the circle into which the path of the ray is coiled depends upon three things—

(1) The strength of the magnet, which is easily measured.

(2) The momentum of the electron—that is, the product of its velocity and its mass or inertia.

(3) The magnitude of the charge of the electron.

Now the path of a cathode-ray is affected also by the presence of an electrically charged object. Suppose a cathode particle travels horizontally between two horizontal parallel plates, the upper of which is negatively charged, so that the electron moves away from it, and the lower of which is positively charged, so that the electron moves towards it, a new downward motion is given to the particle. The path of the ray will now be a curved path, precisely similar to that of a bullet fired horizontally from a cliff. In both cases the particle moving originally uniformly in a horizontal direction tends to fall with a uniform acceleration. That is to say, the path will curve downward in a parabola, and the rapidity with which the particle falls will again depend on three things—

(1) The strength of the charges on the

plates and their distance apart, which are easily measured.

(2) The kinetic energy of the particle—that is, the product of half the mass into the square of the velocity.

(3) The magnitude of the charge of the electron.

From experiments of this character Sir J. J. Thomson deduced the separate values of the mass of the electron, of its charge, and of its velocity. He found that the mass was less than 1/1000th of that of the hydrogen atom. The most recent determinations make it about 1/2500th of the mass of the hydrogen atom.

The velocity varies a good deal with the state of the vacuum in the tube in which the cathode-ray is generated, but does not usually exceed about one-fifth that of light or 37,000 miles a second. Radium, on the other hand, emits β-rays, which are these same electrons travelling free, with a velocity very much greater, and some travel practically with the speed of light itself. The diameter of the electron has been also approximately calculated and found to be only about 1/100,000th of that of an average molecule, so that both in mass or inertia and in size the electron is incomparably smaller than the smallest particle previously known to science. Sir Oliver Lodge has likened the electrons in an atom to flies in a vast cathedral. From being a rare phenomenon, first studied free in a good vacuum, the electron has been found to play a large part in the most varied phenomena, usually associated with atoms of matter. The radioactive substances, however, as has been mentioned, emit β-rays, which are nothing but cathode-rays, or free-flying electrons, only travelling at much higher velocity. In spite of these extensions of our knowledge, only one kind of electron is known, and that is identical in all respects with that described.

Long before the electron was known, however, it had been mathematically predicted that an electrically charged piece of matter must have a greater inertia, and therefore, in this sense, a greater mass than the same matter without the charge. But if we deal with visible charged objects, or even with charged single molecules or ions, the part of the mass contributed by the charge is such a minute proportion of the whole that the prediction had only an academical interest. With the advent of the electron, a particle probably 100,000 times smaller and 2000 times lighter than the smallest hitherto recognised, the prediction became of supreme importance. For it was recognised that the inertia, or mass, of an electron might be entirely due to the perfectly well-known electro-magnetic phenomena which take place when an electric charge is put in motion, and the subsequent development of the subject has borne out this idea. In this one case we have a sufficient philosophical explanation of the property we call inertia, and at once the question arises: Is the unexplained inertia of matter a different thing from the elucidated inertia of electricity, or is it possible that the inertia of matter is due to the same phenomena as that of electricity, and that matter is in some unknown way compounded entirely out of electrons?

We have seen that a current flowing along a wire may be considered simply as a colleetion of electrons moving in the wire from its negative to its positive end. If such a wire is flexible enough it will behave to a magnet exactly like the pencil of cathode-rays we have just considered, and will coil itself round the magnetic lines in circles or spirals. If the wire is coiled up into a spiral, or solenoid, it becomes itself a magnet as soon and for as long as the current flows, but if the current is stopped the magnetism disappears. The magnetism is enormously strengthened, but not otherwise changed, if a bar of iron is inserted into the coil, thus forming what is known as an electro-magnet. From such a solenoid, with or without its iron core, magnetic lines extend from the ends just as in the case of a bar magnet. The surrounding space has been changed, in that it has become magnetised by the motion of the electrons in the wire. It is impossible without a knowledge of electro-magnetism to go into these matters very deeply, but it will be sufficient if we confine ourselves to one important general aspect of this change which takes place when the current flows in the solenoid and the surrounding space becomes magnetised. When the current is flowing the surrounding space has been endowed with energy, known as electro-magnetic energy, which it does not possess when the current does not flow. So we arrive at the philosophical explanation of the inertia of an electron. The space around an electron at rest becomes endowed with energy when the electron moves, and before the electron can again be stopped this energy must be withdrawn. This we have seen is what we mean when we speak of the attribute inertia.

We have seen that a stream of electrons flowing at right angles across a magnetic line experiences a sideways thrust, in a direction at right angles both to its original path and to the magnetic line of force. Instead of moving the electrons past the magnet, we may move the magnet past the stationary electrons, which will experience a thrust just as before, and if free to move—if, for example, they are the electrons resident in a piece of metal forming part of a closed metal circuit—we shall get a flow of electrons in the metal in the stated direction. That is to say, we may generate an electric current by the passage of a wire through the poles of a magnet. Imagine a straight copper wire stretched horizontally, and a magnet moved from above vertically downward so that the wire passes between the poles cutting the magnetic lines between them. The electrons in the wire will be urged along it as the magnet passes, and if the ends of the wire are connected to any circuit, through which a current is capable of flowing, a momentary current will traverse the circuit. The modern dynamo works on this principle, and may be looked upon as an electron pump. The wire is arranged to lie parallel to the axis on the surface of a cylinder spinning between the poles of a magnet, and so in its revolution is continually crossing and recrossing the magnetic lines. Usually a large number of copper wires wound in a peculiar way on an iron core—the whole being termed an armature—rotates between the poles of a powerful stationary electro-magnet. The thrusts which the electrons in the wire experience as the wires cut the magnetic lines are usually arranged, by means of a commutator, to act all in the same direction and so all unite to drive the electrons out of one pole and to draw them in at the other pole of the machine. When the outer circuit is open, so that no current flows, the power necessary to drive the machine is only that wasted in friction of the bearings, etc. But when the circuit is closed, so that a certain current flows, an added amount of power is necessary to force the electrons against the resistance of the circuit. The dynamo transforms this added mechanical energy—the kinetic energy of moving matter—nearly quantitatively into electric energy—the kinetic energy of moving electrons.

If now with the dynamo at rest, but with its magnet excited, we pass a current through the armature, the machine becomes an electric motor. The electrons now urged by outside energy through the wires in the magnetic field turn the wires and with them the armature. Perfect reciprocity is the feature of these mechanical, electric, and magnetic actions. Very few mechanical engines are reversible in this sense. If you drive a steam engine you do not raise steam in the boiler. In the world of electricity and the ether, however, nothing can move in one dimension of space without attendant consequences in the other two dimensions.

These relations between the electron and the external field of energy which attends its motion are perfectly reciprocal. On the one hand, the electron cannot move from rest without this attendant field of energy around it coming into existence, and cannot be stopped without this attendant field of energy disappearing from the ether. On the other, the creation of a magnetic field in space endows, or tends to endow, all the electrons in that space with motion in one direction, and the cessation of the magnetic field endows, or tends to endow, them with motion in the opposite direction. A steady current of electricity, or flow of electrons in a circuit, produces no change in the surrounding magnetic field. A steady magnetic field produces no movement of electrons within it, and so no electric current. Electrons changing the speed or direction of their motion produce attendant changes in the surrounding magnetic field, and conversely a change in the direction or strength of a magnetic field produces motion or tendency to motion of the contained electrons. These phenomena, well known since the discoveries of Faraday, who termed them electro-magnetic induction, now lie at the very foundation of the modern science of electrical engineering. They have no mechanical analogies, for they belong to a more fundamental world than that of matter. It has been stated that space has three dimensions—length, breadth, and depth—rather than any of the other numbers which mathematicians have attempted to picture, because of the peculiarities of the ether, in which motion in any direction is attended simultaneously by influences acting in the two directions at right angles to it and to each other. However this may be, until it is possible to educate the mind so that it apprehends intuitively the three dimensional aspects of motion in the ether, the electro-magnetic world, which underlies the material world and which may completely embrace it, must remain a foreign element as difficult to breathe as air is to a fish, by those accustomed only to the grosser ideas of matter and its motion. It is a time of transition. The discovery of the electron has to a certain extent rendered the subject concrete and picturable, but the pioneers even have hardly yet cleared their way through the jungle of obsolete and confusing habits of thought which naturally still surround the subject. Only a few of the more important cases can here be attempted, and the first of these is one of the oldest.

CHAPTER VIII

RADIATION

THERE is still a gap in the chain of reasoning to be supplied. When an electron is being urged on to move from rest at a continually increasing speed, surrounding its path there is being built up, as the expression goes, a magnetic field of greater and greater strength. The energy being put into the electron overcoming its inertia flows out continuously into the space surrounding its path and travels along with it. When the electron is being retarded and brought to rest, the energy it, itself, is now supplying, by virtue of its inertia, flows in from the surrounding space and the magnetic field gradually weakens and disappears. As we know that the magnetic field extends for a considerable distance all round an electromagnet, these outgoings and incomings of energy, between the electron and its environment, involve the transference of energy over considerable distances, and as the air or surrounding matter plays no part in the phenomena, they must take place through empty space. It is not too much to say that the idea of an *ether* has been invented by scientific men for the express purpose of accounting for the flow of energy across empty space, and is at present little more than a term to express the medium in which these transferences occur. Action at a distance, be it gravitation, electric pellation or tractation, or magnetic action, carries with it the necessity of supposing something to exist in the intervening space, and science takes the simplest possible view when it supposes that one such universal medium, the ether, fills all space, alike between the molecules of a piece of matter as between the most distant stars, and that in this one medium all these various influences are transmitted. Lest, however, it be supposed that the ether is purely a philosophical way of escape from the unknown, and that we know nothing of this elusive and all-pervading medium, let us pause for a moment to consider the origin of the energy which animates nearly everything that moves on this earth. The train that rushes on its journey bearing its hundreds of tons of weight 1000 miles in the day, the liner bearing its tens of thousands of tons lightly across the seas, are animated with the energy that reached this earth, from a place 90,000,000 miles away, as sunbeams during the forgotten ages of the past. Almost all the energy, with which the modern world throbs, arrived through this medium which connects us with the stars, and the thought on which we stumbled, considering the outflowings and incomings of energy around a moving charge, or current of electricity, underlies the flow of all energy throughout the universe. The luminiferous ether, as it was first called, because it bore the light across intra-stellar space, has at least one very definite characteristic, which is quite sufficient to entitle it to be considered as a physical reality. It transmits light and, so far as is known, every other influence which traverses it, at a definite speed—185,000 miles in the second. This velocity is as characteristic of it as the velocity of sound in air, 1200 feet per second, is characteristic of the atmosphere. Radiations, be they light or heat, whatever their colour or wave-length, X-rays, the ether-waves employed in wireless telegraphy, magnetic disturbances, whether they reach us from the sun as the accompaniment of solar storms, or whether, lastly, they circulate around the space surrounding a wire in which current of electricity is being started or stopped—all travel through space with the speed of light. Sound is the vibration of the air, and all the gamut of sounds and noises are essentially air disturbances of the same type. Radiation is the vibration of the ether, and all the various phenomena just enumerated are due to electro-magnetic changes accompanying the alteration either of the speed or direction of motion of electrons. The ether, so far as we know, vibrates only in this one way, and the vibrations are transmitted only with one velocity. The explanation of the inertia of the electron embraces also the phenomena of radiation in all its numerous forms.

Electrons are a normal constituent of

matter, and are found revolving round the atoms much as the planets of a solar system do around the central sun. The electrical tractation between the electron and the positively charged atom takes the place of gravitation. The electrons, like the planets, possess a period of revolution appropriate to their distance from the atom. The nearer they are the faster they must revolve, the law regulating the period of revolution and diameter of the orbit being completely analogous to Kepler's law for the planets. But owing to the minuteness of the atom and the relative greatness of the electrical tractation compared with gravitation, the periods of revolution of electrons are almost inconceivably short. These electrons revolve thousands of millions of millions of times per second, so that it might be supposed that it would be quite impossible to measure them. As a matter of fact, no magnitudes in science are known with greater exactitude.

Consider the electron revolving round the atom. At any point of its revolution an electro-magnetic field of energy extends around it, appropriate to its motion. When it has traversed half a revolution it is moving at the same speed as before, but in exactly the opposite direction. Its attendant magnetic field is therefore exactly reversed in direction. The reversal is not a sudden but a rhythmic process occurring twice in the revolution. At any point in space the magnetic field attains a maximum, diminishes to zero, reverses its direction, and attains a maximum again equal and opposite to the former maximum in half a complete revolution. The ether transmits these rhythmatic changes in the field of energy surrounding a revolving electron outwards through space, as always, at its own peculiar velocity, the velocity of light. And these rhythmic movements of the ether, produced by the smallest entity known reversing its direction of motion regularly in its orbit round the relatively massive single atoms of matter thousands of billions of times in the second, what inconceivably delicate instrument can it be that science has invented for their detection and study? Ah! no instrument maker can make such an instrument. It is furnished ready made not only to man but to some of the lowliest organisms that inhabit the world. The phenomena being described is radiation and the rhythmatic vibrations of the ether, if they occur within certain limits of frequency, constitute light.

Instead of speaking of the periods of revolution, it is more usual to speak of the wave-lengths. The velocity of transmission is always the same, 185,000 miles per second, so that since in one second the number of complete wave-lengths is the same as the period of revolution, and must cover a distance of 185,000 miles when placed end to end, it follows that the length of the complete wave is the velocity of light divided by the period. If the period is rapid the wave-length is short, and *vice versa.* For deep red-light, at one end of the visible scale of the spectrum, the wave-length is about 7/10,000th of a millimetre (1 mm. = 1/25 inch); for deep violet light, at the other end, about 4/10,000th mm. Waves shorter than this in the ultra-violet, down to 2/10,000th mm., affect the ordinary photographic plate, but not the eye. To these rays glass is opaque, but quartz and fluorspar remain transparent, whilst for still shorter waves even the air and gases in general are no longer transparent, but absorb the rays, becoming "ionised" as by the X-rays. So that still shorter rays can only be studied in a vacuum, and fluorite is the only transparent optical material available. These ultra-violet radiations are capable of producing brilliant fluorescence,—for example, in barium platinocyanide. The retina of the eye fluoresces faintly also, and this accounts for the fact that far beyond the limit of the extreme violet of the spectrum, lines can be seen with a faint neutral or lavender tint. Longer waves up to 20/10,000th mm., in the infra-red, can also be studied by photography with specially prepared emulsions. The ordinary emulsion is of course not affected by red or infra-red light. Rock salt is transparent to these long rays, but glass is more or less opaque, whilst ebonite in thin sheets is transparent even in the visible red region. If the sun is viewed through an ordinary camera shutter, made of thin sheet ebonite, it may be distinctly seen as a faint deep red object. There is one generalisation that can be made about transparent solid substances. They are all fine electrical insulators. Mica, ebonite, glass, quartz, amber, sulphur, etc. are among the most perfect insulators known, and all exist in transparent or translucent forms. Metals and good conductors of electricity are opaque. Moreover, in those cases where it is possible to get a transparent film of metal, as can be done with gold and silver leaf by heating, the transparent form of the metal is found to have lost its electric conductivity.

This, no doubt, is connected with the fact that in insulators the electrons are anchored to the molecules and can vibrate with them but cannot move about, so that when the electromagnetic wave of light crosses them the electrons vibrate in unison with the light, but the energy is not dissipated in moving them bodily.

The distinction between radiant heat and light is non-existent. In being absorbed by opaque objects, radiations of all wavelengths, whether they belong to the visible or invisible region of the spectrum, are transformed into heat. Very small amounts of radiant energy in the green and yellow regions are visible to the eye, and in the violet and ultra-violet can affect the photographic plate, whereas our experience of the longer waves is usually confined to sufficient quantities to be detected by their heating effects, as radiant heat. In addition, when a solid substance is gradually heated, its radiation is confined to these dark infra-red or heat rays, and not until its temperature is raised to "red-heat," about 500° to 600° C., does it begin to emit any visible light. In passing to the temperature of a white heat, not only are waves of shorter and shorter light emitted as the temperature rises, but also the radiation of the longer waves is enormously increased at the same time, so that even white-hot bodies give out far more energy as dark heat-rays than as visible light-rays. In astronomy the distance to which the spectrum of a star extends into the violet region is used as a measure of the temperature of the star. Even the eye distinguishes red or comparatively cool stars like Betelgeuze from blue or hot stars like Vega. For every temperature of a heated substance, there is one particular wave-length of which more radiant energy is emitted than of any other. When glowing substances of higher and higher temperature are examined, it is found that the maximum emission of energy shifts regularly farther and farther along the infra-red region towards the visible region of the spectrum.

An extraordinary case occurs in this connection of what the unsophisticated person, who commented upon the fact that large towns always occur on navigable rivers, would term Providence. The sun is probably a red or comparatively cool star, but its temperature, estimated at 6500° Centigrade, is incomparably greater than the highest, about 3500°, attainable on earth even in the electric furnace. For such high temperatures the maximum amount of radiant energy is no longer situated in the infra-red or dark heat region, but is for the solar temperature in the visible region between the yellow and the green, at the point, that is to say, of maximum sensitiveness of our eyes. What can this mean, but that the human eye has adapted itself through the ages to the peculiarities of the sun's light, so as to make the most of that wave-length of which there is most. For a star cooler than the sun the maximum of energy would be toward the red, for one hotter than the sun toward the violet. Hence if these suns have planets peopled with inhabitants, their eyes, if they adapt themselves, as ours seem to have done, to suit the prevailing conditions, would be most sensitive to red or violet respectively, and the yellow green of the spectrum which appears so vivid to us would be to them relatively dull. Let us indulge for a moment in those gloomy prognostications, as to the consequences to this earth of the cooling of the sun with the lapse of ages, which used to be in vogue, but which radioactivity has so rudely shaken. Picture the fate of the world when the sun has become a dull red-hot ball, or even when it has cooled so far that it would no longer emit light to us.

That does not at all mean that the world would be in inky darkness and that the sun would not emit light to the people then inhabiting this world, if any had survived and could keep themselves from freezing. To such, if the eye continued to adapt itself to the changing conditions, our blues and violets would be ultra-violet and invisible, but our dark-heat would be light, and hot bodies would be luminous to them which would be dark to us. One can hardly emerge from such thoughts without an intuition that, in spite of all, the universal Life Principle, which makes this world a teeming hive, may not be at the sport of every physical condition, may not be entirely confined to a temperature between freezing and boiling-point, to an oxygen atmosphere, to the most favourably situated planet of a sun at the right degree of incandescence, as we are almost forced by our experience of life to conclude. Possibly the Great Organiser can operate under conditions, where we could not for an instant survive, to produce beings we should not, without a special education, recognise as being alive like ourselves.

The mathematician's way of expressing a change of velocity is to say that the velocity is accelerated, and this strictly scientific use of the term *acceleration* includes the stopping or retardation of motion, and the change of direction of motion, as well as the mere increase of speed signified by the current use of the word in ordinary speech. Radiation is the consequence of the acceleration of an electron in the scientific sense. An electron revolving round an atom, like a planet round a sun, experiences a constant acceleration towards the centre, and the radiation is rhythmatic and regular so long as the orbit of the electron remains the same, thousands of billions of precisely similar waves following each other out into space every second. We have, however, already had to consider a far simpler case than this. The electrons of the highly exhausted X-ray tube suddenly, in full flight, strike an obstacle made of the densest possible material. Their course is suddenly arrested and they are brought to rest. Let us fix our attention on a single electron in full flight. We know that surrounding it there is the appropriate field of energy. The next instant the electron has struck the anti-cathode and has been brought to rest. The field of energy around it suddenly changes. The ether, however, cannot instantaneously transmit this change. If a long straight line of soldiers in open formation is sweeping regularly across a plain and an officer at one end cries "Halt!" the soldier next to him halts instantly at the word of command, if he is an "ideal" soldier, that is to say, and possesses no inertia. But the man at the other end of the line, say three hundred yards away, however instantaneously he obeys the word of command when he receives it, does not receive it till about one second after it has been given, which is the time taken for the sound to travel. In reviews of troops in open formation the sound wave may, as it were, be seen distinctly travelling along the line. In exactly the same way the order radiates from the suddenly arrested electron, outward through space with the velocity of light in single wave or pulse. But there is no rhythmal or periodic succession of waves as in light. The bombardment of the anti-cathode by the electrons which produces the X-ray is, compared with light, like the noise of the patter of hail on a roof compared with a musical note of sound. The next illustration of the radiation attending the acceleration of electrons is the Hertz waves used in wireless telegraphy, but space forbids their detailed consideration.

CHAPTER IX

RADIOACTIVITY

UNTIL 1896, the observed facts of science were in agreement with the view that the atom was the ultimate limit of material subdivision, and that in no known changes, chemical or physical, including in the latter term electrical and electro-chemical, did the atoms belie their designation as the ultimate foundation-stones out of which the whole material universe was built up. Negative electrons came to be recognised as particles smaller than the atom, and opinions were expressed that in some unknown way atoms were compounded out of these electrons. The philosophical explanation of the inertia, or mass, of electrons raised the question whether the mass or inertia of matter was essentially different from that of electricity. The possibility was present, therefore, that, if by some means the overpowering pellation of electrons could be neutralised, and a very great number, many thousands or hundreds of thousands as the case may be, could be crowded together into the space occupied by an atom, that might be the atom and matter as a separate existence might be referred to a condensed form of electricity. The progress of science has moved away from this simple conception. Positive electrons, which were postulated as the "cement," whereby the pellation of the negative electrons might be neutralised, has remained merely a term to explain the supposed condensation of electricity into matter, and has as yet no physical or experimental basis of existence. In addition two sorts even of negative electrons have had to be postulated. The one, free electrons, which move freely among the atoms of matter, can be withdrawn from or added to atoms, without the necessity of supposing that the atom, as a separate entity, has thereby been essentially altered, or has ceased to exist as such. The other kind consists of the purely hypothetical "structural electrons" out of which the atoms themselves, by hypothesis, are really built up. The electric charges which make their appearance on the rubber and the object rubbed in frictional electricity, it would be altogether far-fetched to regard as derived from the disappearance from existence of the equivalent amount of matter. Gradually *all* the known phenomena due to electrons, even the lines of the spectra of elements, which at one time it was thought revealed its most intimate internal construction, have been associated with the first kind of electron, which pursue a joint existence with the atoms, much as the attendant satellites or planets do in reference to their central suns. The second class, in other words, the atoms themselves, remain, as they were, untouched by these advances. There are still eighty or more distinct types of elements, each with a specific type of atom or smallest particle, which as yet can neither be expressed in terms of one another, nor of anything more fundamental. On the other hand, a totally distinct experimental science, radioactivity, has grown up since 1896, which derives first hand from Nature, most important and astonishing evidence of the properties of these atoms, which till then had been entirely unsuspected and unpredicted by the theories of physical science. Events have proved that chemistry is not the most fundamental knowable science of matter, and that changes are proceeding slowly and spontaneously in certain *atoms*, those of the elements exhibiting the new property of radioactivity, which are totally distinct from the kinds of change which have hitherto been studied.

The discovery of the property of radioactivity, by Becquerel in Paris in 1896, was, experimentally, closely related to that of the X-rays by Röntgen in 1905. Becquerel examined certain fluorescent substances, that is, substances which have the power of absorbing light and other radiations and of re-emitting it, changed to a colour characteristic of the fluorescent substance, rather than of the kind of radiation by which it is produced. By good fortune he included certain of the salts of uranium, which fluoresce with a beautiful greenish yellow hue. He so discovered that these substances emit also new kinds of rays, which, like the X-rays, traverse opaque substances like cardboard and thin metal foil and affect the photographic plate. Continuing, he proved that the new rays had nothing to do with the property of fluorescence, but were a constant entirely new property of the element uranium, exhibited under all circumstances in unaltering degree by all its compounds and by the element itself. Uranium is distinguished among the elements as the last member of the Periodic System. It has the heaviest atom of all the elements and its atomic weight is 238.5, when that of oxygen is taken at 16. Mme. Curie made an examination of all the known elements or their compounds, to see if this new property was possessed by any others than uranium. She found that thorium, the next heaviest element, with atomic weight 232.5, possesses a similar property. None other of the known elements are radioactive. But the natural ores of uranium, the minerals such as pitchblende, in which this element is found in the earth, possess a greater degree of radioactivity than can be accounted for by the uranium therein contained. The same has since been found true for the thorium minerals. She proved that part only of the radioactivity is due to uranium, and that other new radioactive elements, in excessively minute quantity, are present. One of these, happily given the name "radium," after many years of patient work was separated from the mineral, and its compounds were prepared in the pure state. Its atomic weight proved to be 226, the next highest to thorium, and in its whole chemical nature it was just what might have been expected from the Periodic Classification of an element with this atomic weight. It resembles very closely barium and the other members of this family, strontium and calcium. The compounds of this group of elements are well known; for example, lime is the oxide of calcium, but the elements themselves are very reactive metals which are difficult to isolate from their compounds. Last year, however, she succeeded in isolating the element radium, and it proved to be a metal very similar to barium, so far as its mere chemical nature is concerned. The amount of radium in good pitchblende is little more than one part in ten million. Ten tons of pitchblende thus contain only 1 gram of pure radium, and in spite of the interest awakened, the total amount of radium that has ever been prepared probably does not exceed 10 grams or about $\frac{1}{3}$rd of an ounce. Yet, even in pitchblende, the radioactivity contributed by the radium greatly exceeds that due to the uranium. The pure radium compounds are, weight for weight, many millions of times more radioactive than those of uranium or thorium, and many remarkable properties are exhibited by them which uranium and thorium, on account of the feebleness of their radioactivity, do not show. Radium gives a characteristic spectrum distinct from any other known substance, and connected, by certain mathematical relations between the wave-lengths of its lines, with the spectra given by barium, strontium, and calcium. This is of the highest importance, for it establishes the title of radium to be considered a new element beyond all question. If radium is no true element, then the word "element" has no meaning.

At the outset it may make the matter clearer if it is stated that the chemistry of the radio-elements, uranium, thorium, radium, etc., is in no way exceptional, but that, superimposed upon their chemical properties and totally unconnected with them, the elements exhibit an entirely new set of properties, which may be termed the radioactive properties. The radioactivity is a property of the atom, and neither the particular compound in which the atom is combined, the physical state or conditions, such as temperature, concentration, etc., nor the past history of the substance, have any real influence upon it.

In considering these radioactive properties, the nature of the rays, emitted by the radio-elements, first calls for remark. In this department the pioneer was Rutherford. The general methods of studying the new radiations are similar to those employed for the X-rays. First, the rays affect the photographic plate; secondly, they, when sufficiently intense, excite visible fluorescence in the well-known fluorescent substances already described; and lastly, they "ionise" the air. The air is normally an almost perfect insulator. A gold leaf electroscope, when charged, retains its charge for hours or days, in spite of the fact that it is being bombarded incessantly by the countless molecules of the air. But if traversed by X-rays or by any of the new rays, even also by light of exceptionally short wave-length, the air absorbs the rays and suffers a change. The neutral molecules are dissociated into oppositely charged *ions*, and these ions carry the electricity through the air, so that it becomes a partial conductor. A gold-leaf electroscope is discharged by X-rays and the rays from radioactive substances, and this property has proved of the utmost service, for upon it an accurate system of measurement has been based. It may be stated that any of the new rays are easier to deal with quantitatively than common light is, because their intensity is readily and accurately measurable by electrical instruments, like the gold leaf electroscope.

Three kinds of rays are distinguishable, termed Alpha or α-rays, Beta or β-rays, and Gamma or γ-rays. The β-rays are the ones most obvious on first examination, for they affect the photographic plate powerfully and are capable of traversing metal foils. Their penetrating power is somewhat less than that of the average X-rays, but it is sufficient to be remarkable. The γ-rays are very feeble by comparison, and very active preparations are necessary to exhibit them. But their penetrating power is by far the greatest of any known kind of ray. The γ-rays of radium traverse half an inch of lead before being half-absorbed, and other substances, roughly in proportion to their density. The fluorescent effects of the β- and γ-rays are best shown with willemite and the platinocyanides. The α-rays are among the most feebly penetrating of the new kinds of radiation, and are absorbed by a single sheet of paper or by a few inches of air. Nevertheless they are by far the most important class, and possess over 95% of the energy evolved from radioactive substances. They produce very powerful ionising action and also brilliant fluorescence in zinc sulphide, diamond, etc., but their photographic action is relatively feeble, and their effect on the fluorescers, which show best the β- and γ-rays, is small.

The β- and γ-rays are believed to stand in a relationship similar to that of the cathode-rays and X-rays already discussed. The β-rays are free-flying single negative electrons, but their velocity is, in some cases, almost that of light itself, the fastest velocity known. They are deviated by a magnet just like the cathode-rays, but less easily on account of their much greater kinetic energy.

When these rays impinge upon atoms of matter, they are all, more or less, according to their nature, absorbed, and their kinetic energy, as always, is transferred to the molecules of matter and becomes heat. The heat so generated by pure radium compounds is extraordinary, considering the minute quantities of radium which are available. Every hour radium generates sufficient heat to raise the temperature of its own weight of water from the freezing-point to the boiling-point. In one day the energy generated is sufficient to decompose its own weight of carbon dioxide into carbon and oxygen, and, in thirty-eight hours, its own weight of water into hydrogen and oxygen. These are among the most energetic chemical reactions known. Yet, year after year, since the substance was discovered, radium has been pouring forth this steady stream of energy and shows no sign of failing. In ten years the energy generated equals that developed in the combustion of over a thousand times its weight of pure carbon, and more than this of any ordinary fuel. Yet these supplies of energy continue unabated and unaffected by any considerations whatever. There is no way of turning the stream on and off as it is wanted. This property of continuously evolving energy is as much an essential part of the nature of radium as their unalterability in the fire is a property of the noble metals.

How are these discoveries to be reconciled with the law of the conservation of energy, and with the view that energy is a definite physical existence which must come from somewhere if continually generated? They have been reconciled completely, but the explanation involves the view that the atom of the chemist, although still the ultimate limit of subdivision of matter in every artificially engendered process, is not the natural limit. This explanation was put forward ten years ago by Rutherford and the writer, and has since been adopted. In the naturally occurring phenomenon of radioactivity there is a spontaneous process continuously going on, in which the atoms themselves are the units that change. The oft-quoted words of Clerk Maxwell, before the British Association in 1873, are no longer true. He said: "Natural causes, as we know, are at work which tend to modify, if they do not at length destroy, all the arrangements and dimensions of the earth and the whole solar system. But though in the course of ages catastrophes have occurred and may yet occur in the heavens, though ancient systems may be dissolved and new systems evolved out of their ruins, the molecules out of which these systems are built—the foundation-stones of the material universe—remain unbroken and unworn." In present-day nomenclature the word "atoms" must be substituted for "molecules" in this quotation, for before the discovery of radioactivity even the most eminent physicists had not, like chemists, learned clearly to distinguish between atoms and molecules. The dissolution of ancient systems and the evolution of new ones out of their ruins, referred to so eloquently by Clerk Maxwell, are in all probability controlled by the dissolution of their atoms. The infinitely slow march of cosmical evolution probably keeps pace with the gigantic periods of time which these atomic changes require.

If radium, the element, is the source of the energy it pours out in a continuous stream, radium, the element, must change, and the change of an element is transmutation. The position of the parts constituting the atom must alter, and the energy associated with the atom must suffer conversion into the energy of motion, which ultimately appears and can be measured as heat. Evidence of these changes was soon forthcoming, and now, complicated as some of them are, every detail almost of the process, whereby the energy is evolved, is known to a degree of accuracy unsurpassed in many of the older examples of material change. Several reasons exist for this. Radioactivity is an inevitable process, which is quite independent of the conditions and circumstances, and indeed is not known to be really affected in the slightest degree by any circumstance whatever. Whereas chemical changes, notoriously, are far less simple, and are affected and sometimes reversed by a great variety of conditions, many of which are still only very imperfectly understood. In the second place, the electrical methods of measurement employed are of unsurpassed delicacy and certainty, and the changes occurring in a quantity of radioactive matter, which in a few minutes or seconds is easily detectable by these methods, might have to continue for geological epochs of time before they produced any effect that could be detected by the ordinary methods of chemistry. But the most important reason of all is that the changes of radium and the other radio-elements are, on the one hand, neither gradual changes of the atom in which all the atoms slowly evolve their energy and slowly change into new forms, nor, on the other, are they completely sudden processes in which the individual atoms give up their store of energy, changing at one step into the product or products. The changes in radioactivity are of the latter type exclusively, but they are not single. If the atoms of radium changed suddenly into their final products evolving their store of energy in one stage, such a process would be difficult to identify. True, the energy would sufficiently indicate the change, but it would remain the sole indication. The proportion of the radium changing in a year or in ten years is altogether too small to be detectable by ordinary methods with the minute quantities so far available, whilst in uranium and thorium, the radioactivity of which is millions of times feebler than that of radium, the rate of the change is correspondingly smaller. This is of reality a necessity, as a little consideration will show. Radioactivity is a natural spontaneous process occurring in known materials at a constant rate. The earth has existed, according to geological evidence, for hundreds of millions of years in much the same state as at present, and if radioactivity is a change occurring in certain elements, these elements must long ago have disappeared from the earth altogether, unless the changes were slow even as compared with the progress of geological time. There is only one way by which such changes could come within the range of experimental science, and that is the way in which the radio-elements actually have been proved to be changing.

There is usually a long succession of separate sudden changes, part of the energy being evolved at each change. In consequence there exist, intermediate between the initial element, radium, for example, and its final product, the element lead, as is generally supposed but not yet proved, a number of intermediate forms of matter having an existence more or less transitory, but, in spite of their infinitesimal quantity, evolving so much energy in their further changes that they can readily be detected and accurately studied. Nearly thirty of these new transitional forms of radioactive matter have been recognised in the changes of the elements uranium, thorium, radium, and actinium, and considerations of space alone would prevent the detailed consideration of the enormous number of important investigations that have been published upon them in the last ten years. It must suffice to take one example, the first change of radium itself, in some detail, as all the others are strictly analogous in their general character.

The radioactivity of a radium compound appears to consist under ordinary circumstances of all three types of rays in unchanging proportions. It is sufficient to dissolve the compound in water and to evaporate the solution to dryness again, or even, more simply, strongly to heat the compound without dissolving it, to remove by far the greater part of this radioactivity. A few hours after this treatment, the activity of the radium is at a minimum and no further chemical or physical treatment, however elaborate, further alters it. At this stage the β- and γ-rays have been entirely removed, whilst the α-rays have been diminished to one-fourth of their initial amount. The substance radium has not been at all altered by the process. In the course of time the solid compound of radium or its solution, if kept in an air-tight vessel, recovers the activity it has lost. All the three types of rays are regenerated at characteristic rates, and in a month it is as radioactive as initially. These operations may be repeated with the same result any number of times. A closer examination reveals the fact that during the solution, or heating, a gaseous substance, called the radium emanation, escapes. If arrangements are made to collect this gas, it will be found that, generally speaking, the whole of the radioactivity the radium has lost is possessed by the gas. The gas itself is in almost absolutely infinitesimal quantity. Yet its radioactivity is so powerful that no difficulty whatever is experienced in detecting and working with it, for it may be mixed, if necessary, with air and then dealt with by ordinary methods. In the solid radium compound many of the rays, particularly the powerful α-rays, are absorbed by the material itself, but in the gas they have full scope, and the fluorescent action which the emanation produces on bodies such as zinc sulphide, is remarkably brilliant.

The radium emanation itself shares with the argon gases the property of not entering into chemical combination or being absorbed by any known reagent. It is also condensed to the non-gaseous form at the temperature of liquid air, and in these ways it is possible to separate it from all known substances and to study it pure. Then the extraordinary minuteness of its actual quantity and the power of its radioactivity become evident. From a gram of pure radium the gaseous emanation obtained occupies a volume, measured under standard conditions of temperature and pressure, of only 0.6 of a cubic millimetre, the volume of an ordinary pin's head. Yet the rays from far less than a thousandth part of this quantity will cause zinc sulphide to fluoresce in a way that will be plainly visible in an absolutely dark hall to an audience of a thousand people. Indeed, if one-thousandth of the emanation obtainable from a gram of radium were mixed uniformly with the air of a very large hall, say with 100,000 cubic feet, or over 3 tons by weight, of air, no delicate instrument such as is customarily employed in the measurement of radioactivity could be worked in the hall, and the amount in a single cubic inch of the air could still be detected by a sensitive gold leaf electroscope. The unit adopted in certain scientific work is the amount of emanation produced by one million-millionth of a gram of radium, a quantity which itself has a volume of less than a million-millionth of a cubic millimetre, and weighs a million million times less than an exceptionally delicate chemical balance will turn to. Such a quantity contains less than a million single atoms. It is almost incredible, but nothing in science is better established.

The heat evolved from this emanation of radium is in proportion to its radioactivity. The quantity of emanation derived from 1 gram of radium evolves three-fourths of the total given by the radium, and when it is removed from the radium, the latter only gives one-fourth of what it gave before. Now it is almost incredible in any case that a quantity of less than a cubic millimetre of gas can evolve spontaneously enough heat to raise the temperature of three-fourths of a gram of water from freezing-point to boiling-point in an hour. To obtain a single cubic inch of this gas, measured under standard conditions, would require 26 kilograms of pure radium. This, therefore, is almost an impossibly large quantity practically to obtain, but it is interesting to note that the energy such a quantity would emit would be equal to that of a powerful electric arc lamp. The mystery of the source of the energy of radium is increased a million-fold when the nature of its emanation is studied.

How long can such an unparalleled evolution of energy last?

This is just where the interesting point comes in. The emanation of radium does not last. It is not an apparently permanent source of energy like the radium from which it originates. If the activity of some radium emanation, sealed up in a tube, is examined from day to day, it will be found steadily to decay. In four days its activity is only half the initial. In eight days it is one quarter, and so on. In a month it has all practically disappeared. But while these changes are taking place a concomitant set are going on in the radium from which the emanation was derived. It recovers the activity it lost, when the emanation was removed, just as fast as the activity of the removed emanation decays. If at the end of the month, when the radium has fully regained its activity, it is redissolved in water, *a new quantity of emanation is obtained* just as great as at first. Radium is *producing* the emanation. The emanation of radium is the first product of the change of the radium atom. This emanation in its turn is changing comparatively quickly. The change is complete in a month. One-fourth of the energy is derived from the change of the radium, and three-fourths from the subsequent changes suffered by the emanation.

Thus, when a quantity of radium is observed to be apparently pouring forth in an undiminished stream its rays and its energy, year after year, what is really taking place is not so simple. One-fourth of the α-rays result from the change of some of the radium atoms into atoms of the emanation. Three-fourths of the α-rays and the whole of the β- and γ-rays result in the comparatively rapid changes of the emanation and its products, which it is unnecessary particularly to specify here. The steady outpouring of rays is due to a balance or equilibrium, when as many atoms of emanation are produced per second from the radium as change per second into other substances. Thus the amount of emanation present and the radioactivity of the radium tend to become and remain constant when this equilibrium is attained.

Now it is a simple matter to calculate from the equilibrium amount of emanation in any quantity of radium, and from its rate of change, both of which data can be directly observed, how much emanation is being produced from the radium in a given time, and therefore the rate at which the radium itself is changing in producing this emanation.

These calculations and numerous others from entirely independent data lead to the result, which is probably not more than a few per cent. in error at most, that in one year 1/2500th part of the radium changes. In other words, the average period of life, the time, that is, which a radium atom exists on the average before it changes, is 2500 years. The average period that the atom of emanation exists is only 5.3 days. Rather less than one-fifth of the whole changes per day.

It is now possible to calculate the total amount of energy that a given quantity of radium will evolve during its *complete* change, which, of course, requires thousands of years. We have seen that in thirty-eight hours the energy is equal to that required to decompose its own weight of water into hydrogen and oxygen. In the total life of 2500 years the energy evolved would therefore suffice to decompose over half a million times its weight of water. In this decomposition more energy is required, weight for weight of substance decomposed, than in any other chemical reaction. Or, the result may be expressed by saying that 1 gram of radium evolves in its complete change as much energy as a quarter of a ton of coal does when it combines with the oxygen of the air and burns.

At once arises a question: If the period of average life of radium is only 2500 years, how is it that there is any radium still in existence? Historical records go back to many times this period. Even if the whole world were originally pure radium 100,000 years ago, the quantity now present should be less than that actually contained in the common rocks and soils constituting its crust. The answer will be found by asking how it is that there is any radium emanation in existence, as this substance has a period of average life of only 5.3 days. Just as the quantity of radium emanation is maintained in constant proportion to the radium producing it, so if there existed in the minerals in which radium is found a much more slowly changing element, in the change of which radium was produced, the maintenance of radium would be explained. Now radium is found in uranium minerals. Uranium is radioactive and is therefore changing. But into what? Exact experiments have shown that the quantity of radium in minerals is proportional to the quantity of uranium. There are certain exceptions, but these can be explained by the influences, such as percolating water, to which rocks in the earth are subjected, and which may dissolve certain constituents preferentially. Excluding these few cases, in all minerals there is always about 3,000,000 times as much uranium as radium. Experiments have also been going on for many years to see whether uranium preparations, initially quite free from radium, produce radium in the course of time. These have not yet produced a definite result. It is known that the change of uranium into radium is not direct, but that an intermediate substance, ionium, which does produce radium steadily with the lapse of time, intervenes. The average life of ionium is probably at least a hundred times longer than that of radium. So that direct attempts to establish the production of radium from uranium necessarily require a very long time to give a definite result. There is, however, very little doubt. The constancy of proportion between uranium and radium in minerals can only be accounted for on the view that the former produces the latter. Just as the period of average life of radium can be deduced from that of the radium emanation and the equilibrium proportion between the two substances, so also can the period of uranium be calculated. There are 3,000,000 parts of uranium to one part of radium in minerals. Therefore, it can be proved, the period of average life of uranium is 3,000,000 times that of radium, or 7500 million years. But this gigantic period agrees very well with the extremely feeble radioactivity of uranium. The latter is many millions of times less than that of radium, and therefore the substance must be changing many millions of times more slowly. Inordinately long as this period seems to be, compared even with those customarily dealt with in geology, it is necessary to explain many aspects of cosmical evolution, as will later be dealt with. The difficulty with the older physicists was to allow geologists sufficiently extended periods of time for the processes they studied. The utmost it used to be possible to allow them, without trespassing the bounds of physical possibility, as it was then understood, the geologists rejected with scorn as utterly insufficient even for a brief portion of recent geological history. That was before these processes of radioactivity were known in which the energy evolved is a quarter of a million or more times greater than in any previously known process. At least there is no difficulty in accounting for the maintenance of radium from uranium for periods of tens of thousands of millions of years, without of necessity being compelled to suppose that the quantity of uranium in the earth is being in some still unknown way maintained. Geological time, even, does not trace back the history of the earth to periods so distant that the quantity of uranium then in existence must have been very appreciably greater than now. That being so, it would be profitless to speculate in the present state of knowledge as to the origin of uranium.

The essential features of these new processes are that the quantities of the elements undergoing change are almost inconceivably minute, so that were it not that the new changes of matter are accompanied by proportionally enormous changes of energy, they would necessarily be quite unknowable. If the change of the matter is rapid, its quantity is excessively minute. If the matter is, like uranium, not altogether scarce, the changes are excessively slow and the radioactivity of the matter proportionately feeble. It is an old world, and any but the slowest primary changes must long ago have run their course. From this point of view the elements of the Periodic Classification are those which have survived because they are stable. At the extreme end, the elements uranium and thorium, though not entirely permanent and stable, are yet changing so slowly that some still survive. In the light of these researches and the known fact that the elements of atomic weight between 220 and 240 are not entirely stable, but are slowly changing, the abrupt end of the Periodic Table with the element uranium, reads like an interrupted record in which the writer in the last few words had succeeded in conveying a hint of the approaching end.

Slow as these new processes are, and infinitesimal as are the actual quantities of the radio-elements undergoing change even in the most favourable cases, yet the ordinary methods of chemistry and spectroscopy have not proved entirely inadequate in confirming them. It may be mentioned that, in the most favourable case, the spectroscope requires a thousand million times greater quantity of matter than can be detected easily, for example, in the case of the radium emanation. The changes in radioactivity are studied by means of the energy evolved. A totally different problem has to be faced when the chemical nature of the matter finally resulting, for example, in the change of radium, is the subject of inquiry. The difference is analogous to that between observing a meteor and trying to find it afterwards, to discover the nature of the materials remaining when its flight is spent. It is, of course, a great help, when the chemical constituents of a substance are under investigation, to have at least some idea of what is likely to be found if looked for. In radioactivity a very simple process of reasoning served to provide this clue. Radioactive changes are slow, but they are continuous. In the minerals in which the radio-elements occur these changes have been in operation for geological epochs of time. Hence the ultimate products must be present as the invariable companions of the radio-elements, and the older the geological formation from which the mineral is derived the greater ought to be the quantity of these products in the mineral.

As soon as the radioactive minerals were examined from this point of view, a very striking fact transpired. These minerals contain the gas helium, though how it got there and why it should occur only in minerals containing the elements uranium and thorium, had remained one of the unexplained mysteries of science. The name helium is derived from the Greek word for "the sun." The most prominent line in the spectrum of helium, a brilliant yellow line very close to the two well-known sodium lines, had been observed in the spectrum of the solar chromosphere long before the element itself was known. Helium was discovered on the earth in certain minerals by Sir William Ramsay before the property of radioactivity was known, and he had noticed that the minerals which contained it contained also the elements uranium and thorium. But helium is one of the family of monatomic gases which are completely inert and do not form compounds. Neither are they absorbed chemically by any substances whatever. Moreover, helium is of all gases the most difficult to liquefy. Its critical temperature, or temperature below which it cannot be liquefied under any pressure, however great, is but a very few degrees from the absolute zero. What then is the origin of this uncondensible, unabsorbable gas in minerals containing uranium and thorium? The answer suggested itself at once as soon as the view that radioactivity was the manifestation of the spontaneous change of elements. It was predicted that helium was one of the unknown ultimate products of the changes. Being formed in the solid mass of the minerals, which are often of a glassy nature, it is unable to escape, and remains imprisoned until the mineral is dissolved or heated. Once liberated it cannot be put back. Then experiments with radium, by Sir William Ramsay and the writer, proved directly that radium is actually producing the gas helium, just as we have seen it is producing the gaseous emanation. Minute quantities, sufficient, however, to be identified beyond any doubt by the spectroscope, were obtained from radium preparations which had been kept for a few months after manufacture. On solution in water this helium was liberated along with the emanation. Then helium was proved to be generated from the emanation. If the latter was prepared pure, by condensation by means of liquid air and other treatment, and sealed up in a spectrum tube, the helium spectrum gradually developed in the tube as the emanation changed. Later direct experiments have established that helium is the product of most of the radioactive elements. It has been so obtained from actinium, uranium, thorium, and polonium.

But from the very first, before these direct

experiments, evidence had been obtained that probably the α-rays consisted of swarms of material atoms, projected with a velocity hitherto never observed. Rutherford was successful in deviating them by a very powerful magnet, and established that the deviation, besides being almost incomparably smaller than that suffered by the β-rays, is in the opposite direction. This indicated that the α-rays, like the β-rays, consist of swarms of free flying charged particles, but that their charge is not negative, like that of the electron, but positive. True to its character, whenever positive electricity is observed, it is not, like the negative variety, alone, but is attached to material atoms. By methods analogous to those discussed for the electron or cathode particle, Rutherford deduced, from the amount the α-ray was deflected by magnetic and electric fields of known strength, their veolocity, their mass, and their charge. After many years' continuous effort, all three data were obtained separately. The velocity is, in the various cases, between 1/15th and 1/20th of that of light. Heretofore the fastest known moving material thing was the meteor, some of which attain a speed of 40 miles a second. This is only 1/300th of that of the α-particles, some of which travel with the speed of 12,000 miles a second. The evidence at first with regard to the mass and the charge enabled only a ratio to be obtained. This indicated that if the charge were a single atomic charge of electricity, or, in other words, if the *a*-particle were analogous to the manovalent hydrogen or silver ions previously discussed, the mass of the particle was twice that of the hydrogen atom. Whereas, if it were a divalent ion, its mass would be four times that of the hydrogen atom. The latter is the atomic mass of helium.

To settle this point, some exceedingly delicate experiments were performed. The methods of measurement were so improved that they were sensitive enough to detect the expulsion of a single *a*-particle, or the disintegration of a single atom of radium. Under suitable conditions it was found possible actually to count the number of *a*-particles expelled from a known quantity of radium in a known time. This last step enabled the charge of the individual *a*-particle to be measured. This proved that the *a*-particle carries two atomic charges. Its mass therefore is 4, the same as the atomic mass of the helium atom. In every known case where *a*-rays are given by radioactive substances, they have been proved to be atoms of helium expelled with very great velocity and kinetic energy. One of the most beautiful demonstrations ever performed finally established the identity of α-particles with helium atoms. The α-particle, although it is, on account of its relatively much greater size, far less penetrating than the β-particle, nevertheless does penetrate a certain very small thickness of matter. It is possible to blow glass so excessively thin that the α-particles can pass through the glass, although the glass film retains to the full its imperviousness to the molecules of a gas. Tubes were constructed of glass of this degree of thinness of the wall, and were fitted into an outer vessel. It was shown that when the tubes were filled with ordinary helium none whatever leaked through into the outer vessel. But when filled with radium emanation, which gives α-particles, these can pass through the glass, and helium was then found in the outer vessel by means of the spectroscope. Thus beyond any possibility of doubt the nature of the α-particle was established.

It is known that three α-particles, or helium atoms, are expelled in the change of a uranium atom into radium, and that in all, including the subsequent numerous changes suffered by the radium, eight α-particles are expelled. The atomic weight of uranium is 238.5, and if we subtract the weight of three α-particles or helium atoms, which is 12, we get the figure 226.5, which is almost exactly the value Mme. Curie found by experiment. If we subtract the weight of eight helium atoms, or 32, from 238.5, we get the figure 206.5. If then in the changes of the uranium atom nothing else except eight helium atoms are expelled (the mass of the β-particles expelled is too small to be of importance), the final product should be an atom of weight 206.5. The nearest element to this is lead, the atomic weight of which is 207.

Now the same kind of evidence which led to the prediction that helium was one of the ultimate products has indicated that lead is produced from uranium. Lead in important quantity is a constant companion of almost all uranium minerals. The older the geological formation from which the mineral is derived, the higher the percentage of lead appears to be. The direct proof or disproof of this view is not yet accomplished, but experiments are now in progress, and the result may be announced at any time. It is generally conceded that there is very little doubt that lead is the final stable form assumed when the changes of uranium are complete.

These investigations, although only in their infancy, have thus thrown a flood of light on the nature of the atom. In the few cases in which it is now possible to watch the process, the outstanding feature is the altogether unparalleled amount of potential energy associated with the atomic structure, which is released and rendered apparent when the structure undergoes change. The element helium seems to play a predominant part in the internal structure of the radio-elements. Uranium appears to be made up of eight atoms of helium with one of lead, and so on. But atoms are not simply a special sort of chemical compound in the ordinary sense. There is more difference between the change of an element, as in radioactivity, and an ordinary chemical change, as in the union of hydrogen and oxygen to form water, than there is between the latter and a so-called physical change, like the condensation of steam. It is best therefore not to be in too great a hurry to abolish the old distinction between atoms and compounds of atoms, or to believe the ancient theory that matter, though apparently diverse, is philosophically simple, and that all the atoms are merely various compounds of some primordial stuff or "protyle." Such a problem has now altogether too serious consequences for the whole future destiny of the race to be lightly assumed, on account of the philosophical edification it arouses. The advance of science along that road, which as a matter of fact never appeared so impassable as it does to-day, would not leave a single stone standing of all the elaborate superstructure of civilisation as it is at present understood.

CHAPTER X

COSMICAL ENERGY

EVER since the doctrine of energy became established the source of cosmical energy has attracted attention. Whence comes the steady supply of solar radiation upon which existence on this planet absolutely depends? How long has it been going on and how long will it continue? Vast as the sun is, something more than its mere store of heat energy is necessary to enable it to continue emitting its light and heat at the present rate for more than a very limited time. Even the idea that the sun is a huge fire and derives its heat from the combustion, or chemical combination of its component materials, is altogether insufficient to supply the waste of heat even over the period covered by the records of history. The idea which has found most favour hitherto is that the maintenance of the heat of the sun is explained by shrinkage of its materials into smaller volume. If, as explained on p. 40 of *Astronomy* in this series, the gravitation has not already packed the materials of the sun together as closely as they will go, and every thousand years the diameter shrinks about 40 miles, the kinetic energy resulting from the gravitation and concentration of the materials would equal that estimated to be lost by radiation. The question at once arises, "Why should the sun shrink unless it is falling in temperature?" The explanation would account for a slower fall of temperature than would occur if shrinkage did not take place, but the idea that shrinkage can maintain the temperature uniform, or nearly uniform, seems inadequate. The heat so gained is, however, altogether insufficient on present evidence, and as there is no other ground for believing that shrinkage is in fact taking place, it is unnecessary to pursue the question.

What has to be accounted for is not the maintenance of the sun's heat at uniform temperature for a few million or even one hundred million years, but for hundreds, possibly thousands of millions. The geologists are agreed, and their point is generally admitted, that the surface of the earth must have remained in much the same physical conditions as at present for these immense periods. Before the discovery of radioactivity, no source of energy sufficiently abundant and lasting was known which would suffice for a period, at the very most, of more than a hundred million years.

Another more definite line of evidence is obtained from the temperature of the earth. The solid crust of the earth conducts heat excessively slowly. If, as was before supposed, the earth was originally hot and molten and has cooled on the surface by radiation, it is possible, by finding the temperature gradient as we descend, that is the number of feet on the average which have to be descended for the temperature to rise one degree, to calculate how great a period of past time could have elapsed before the surface must have been too hot to sustain life. Arguments along these lines put the limits of the possible age of the earth during which it has been in a state capable of supporting life on the surface, as even less than a hundred million years.

These arguments and calculations have all been completely altered in consequence of the discovery of the energy evolved in radioactivity, when atoms undergo spontaneous change. A thousand millionth of a milligram of radium is about the smallest quantity that can be practically detected. It has been found that all the common rocks and soils of which the earth's crust is built up contain measurable amounts of radium. Strutt has calculated from his measurements, that if a crust of the earth only some 50 miles thick contained the same amount of radium as the representative samples of the rocks he examined, the heat generated by this radium would suffice to account for all the heat lost by the earth by radiation. There is, however, no reason to limit the radium to the surface crust. In addition, no allowance was made in the calculation for the heat generated by the small amounts of uranium and thorium, which more recent investigations have shown are as important as the radium in contributing heat. Argument based upon this evidence is, necessarily, somewhat hypothetical, for, under the enormous pressures that exist in the interior, even radioactive changes may not run their normal course. But the day has gone by when the earth is regarded simply as a cooling world. It has in its known material constituents a steady source of fresh heat, which will last, not for one million, but for thousands, or tens of thousands of millions of years. It is regarded as more probable to-day, that instead of the earth becoming colder by radiation, as has been supposed, it is steadily growing hotter and hotter in its interior. The heat so generated throughout the mass, on account of the low conductivity of the rocks and materials forming the crust, only very slowly reaches the surface. At the surface, the heat generated escapes by radiation and maintains the temperature uniform. But the interior is almost completely thermally isolated from the surface, and the temperature within, provided that the composition of the materials is similar to that on the surface, must steadily be increasing. Joly has made some interesting calculations of the inevitable results that must attend such a process. Assuming a quantity of radium, and its corresponding amount of uranium, distributed uniformly throughout the mass of the earth, of two parts of radium per million million, which is less than the average found for surface rocks, this would produce an increase in the temperature of the interior by 1800° C. in a hundred million years. So long as the earth's crust remained solid, this heat would only escape by conduction with extreme slowness. But at some time or other, a world so constituted must explode, when the increasing temperature and pressure within overpowers the strength of the crust. According to the same authority, there is no assurance that such a consummation does not await the future, nor evidence that such has not more than once been an event of the past.

Such considerations give an idea of the importance of radioactivity in cosmical processes. Over periods of time, appropriate to the periods at which these new changes proceed, even minute proportions of uranium and thorium distributed throughout a world, must inevitably affect its whole internal physical condition, apart altogether from the influence of external energy and of the rest of the system of which it forms part. Its history is a function of its diameter, the amount of the radioactive elements contained in its materials and the conductivity of the latter for heat, at least forperiods of thousands of millions of years. The amount of thermal energy it possessed when it originated would, at least in all but exceptional cases, long before this have ceased to exert any influence at all. The primary sources of natural energy, by virtue of which the universe keeps going over immense periods of time, are to be sought not in the great masses of glowing matter dotted about the heavens, nor in their motions under the action of gravity, nor in any of the grosser relations between energy and matter in bulk, but in the individual atoms out of which it is made up. No other source is at once sufficiently abundant and sufficiently lasting, probably, even for a single geological age, the period, that is, since the ocean condensed and rain and rivers began their work of denudation and upbuilding. Only a beginning has so far been made into the study of these new unsuspected forms of energy, but enough is known to make it clear that, whether it be so or not, radioactivity alone, including in that term processes involving atomic transformations, is competent to be regarded as the mainspring of the universe.

Changes of this magnitude, which have

swept aside all the prevailing notions as to the origin, past age, and future destiny of this and other worlds, have not left untouched the questions nearer home concerning human history and destiny. It is not necessary to discuss the question here as to how far the radio-elements are alone in parting with their stores of internal energy, or whether, so far, in radioactivity, merely one type has been recognised of changes which, in greater or less degree, may be proceeding in all matter. From the general similarity of nature of the chemical elements, it can be argued that if one could be artificially transmuted the way would be opened for the transmutation of others. The common elements are merely elements which are not changing, but there is the strongest reason to believe that many of them, if they could be changed into simpler forms, would give out supplies of energy no less remarkable than that furnished by the radio-elements. There is, however, no evidence at present that any other of the common elements except uranium and thorium are capable of spontaneous change, and therefore nothing is known of the energy these other elements may contain. It suffices merely to take stock of the new supplies of energy, in sight, so to speak, without troubling whether the advance of knowledge will disclose more. The conversion of thermal energy into mechanical energy, first practically effected by the invention and perfection of the steam engine, has brought about in a single century more permanent change in the manner of living, and even in the habits of thought of the inhabitants of the world, than any combination of political, social, or personal influences could have effected. It is the mastery of man over Nature, as represented by matter and energy, rather than that of one man over another, or of one race over another, to which histories give such exaggerated predominance, which underlies progress. What, therefore, will be the effect of the discovery that, so far, we have been subsisting on the mere by-products of natural energy, and have remained ignorant even of the existence of the primary supplies in the atoms of matter? That it must exert a profound influence upon every department of human thought need scarcely be stated. The forgotten savage who kindled the first artificial fire little knew the consequences that were to follow. We can form a faint conception of some at least of the consequences which would follow the introduction of the new sources of energy into the practical affairs of the world. But although science has discovered these new sources, it by no means can be taken for granted that their practical utilisation must follow as a matter of course.

A moment's consideration will serve to show that although the present problem of how to release for use at will the energy in uranium, thorium, and radium for practical purposes appears in a new light, it is in reality one of the most ancient problems to which, from the earliest times, the energies of the race have been unceasingly and unsuccessfully directed. The natural processes in which the atomic energy is evolved are necessarily either excessively slow or are shown by compensatingly minute quantities of materials. Along with the discovery that a pound of uranium contains and evolves in its changes the same amount of energy as a hundred tons or more of coal evolves in its combustion, is the knowledge that little more than 1/10,000,000,000 part of this is given out every year. These natural processes must be controlled and made to proceed much more rapidly than they do spontaneously before any of the new sources of energy become of the least use in ordinary engineering. The transformation of the uranium into helium and, presumably, lead must be carried out in an artificial manner before the energy of the process becomes available. But this involves nothing less than the transmutation of one element into others. The new problem is but transmutation, although the alchemists, and others who have attempted to solve it, little guessed what success on their part would have involved. We are no more competent to make use of these supplies of atomic energy than a savage, ignorant of how to kindle a fire, could make use of a steam engine.

Naturally all the resources of modern science have been employed in the attempt to influence radioactive processes, to make them proceed in any way differently than they do naturally, and also to imitate them in non-radioactive materials. These attempts have been signal failures, and indeed from the energy point of view they appear rather like trying to influence the course of a bullet by blowing at it. True to its character as a natural process of transmutation, radioactivity is altogether uninfluenced by external conditions. It proceeds at its natural rate with a detachment from and independence of its environment which qualifies it to be used as an absolute standard of time. Nothing that is known will affect the transmutation of one element into others, and, appropriately enough, nothing known is competent to affect the natural process in those elements which are undergoing spontaneous transmutation. It is a mistake to suppose that it is only a matter of time before science succeeds in this quest. It merely has to be stated that the process of radioactivity involves the expulsion of atoms of helium, with a velocity 300 times greater than ever previously known for any material mass or particle, to make clear how little hope there is for a long time either of controlling or imitating it. One of the most likely processes to affect the rate of transformation of a radioactive substance would be the bombardment to which it is itself subjected by the helium atoms expelled by those of its atoms actually breaking up. This means that the rate of the process should be greater for a concentrated radium preparation than for the same element in a diluted state, mixed with a large quantity of non-radioactive materials. But this has been found not to be the case. It cannot be denied that the only process known at all likely to be competent to affect the stability of an ordinary atom is the change of a neighbouring radioactive atom. The evidence so far, however, justifies the conclusion that artificial transmutation, or what comes to the same thing, the release of the energy associated with the structure of the atoms for practical purposes, has not yet been brought appreciably nearer by the discovery of radioactivity. The weapons of the stone age would be of little use in the working of steel.

After all, what has an atom to fear from the utmost that can be done to it in the laboratory? Has it not been subjected in the laboratory of Nature to temperatures immeasurably higher and to pressures of which science has no conception? Its simple existence is eloquent of its fitness to survive. Not without reason have the atoms been termed the foundation stones of the universe. The title is derived not from the laboratory experience of chemists only, for, by the aid of the spectroscope, the materials of the most distant star can be analysed into their constituent elements. Sun and stars tell the same story. The atoms out of which they are composed are the same, with possibly one or two exceptions, as those found on the earth. It seems almost presumptuous to hope that the atoms which continue to exist unchanged under conditions so transcending any that can be reproduced in the laboratory will ever by any conceivable advance of science be displaced from the proud position they occupy in the economy of Nature.

Through metaphysics first, then through alchemy and chemistry, through physical and astronomical spectroscopy, lastly through radioactivity, science has slowly groped its way to the atom. Through the various ideas of phlogiston, imponderable fluids, attractions, repulsions, affinities, and forces, science has ended with the simple universal conception of energy. The discovery of the relation of the atom to energy within the last decade recalls the strange mediæval myth that the Philosopher's Stone, which had the power of transmuting metals, when discovered would prove also to be the Elixir of Life. Transmutation, the pulling to pieces and putting together of atoms, would render available the primary sources of energy which maintain the time-defying processes of cosmical evolution.

Civilisation, as it is at present, even on the purely physical side, is not a continuous self-supporting movement. The conditions under which it originates determine its period and fix the date of its decline. It becomes possible only after an agelong accumulation of energy, by the supplementing of income out of capital. Its appetite increases by what it feeds on. It reaps what it has not sown, and exhausts, so far, without replenishing. Its raw material is energy, and its product is knowledge. The only knowledge which will justify its existence and postpone the day of reckoning, is the knowledge that will replenish, rather than further diminish, its limited resources. When coal is exhausted and the other physical resources of civilisation have all been squandered, when expanding civilisation is met by a dwindling supply of energy, either science or the atom will have been tested to destruction and one or the other will be the arbiter of the future.

The triumphs of science over Nature till now resemble somewhat schoolboy successes. This period is passing away. Its function in the future will be not the spending to good purpose of what has been provided, but the provision of what is being spent. It is perfectly obvious that, with the whole planet in measurable distance of being occupied, and nations being concerned rather to preserve what they have got than to acquire more, a turning-point is being reached in the upward progress which has hitherto kept pace with the advancement of knowledge. Thoughts of economy and conservation will inevitably replace those of development and progress, and the hopes of the race will centre in the future of science. So far it has been a fair-weather friend. It has been generally misunderstood as creating the wealth that has followed the application of knowledge. Modern science, however, and its synonym, modern civilisation, create nothing, except knowledge. After a hand-to-mouth period of existence, it has come in for and has learned how to *spend* an inheritance it can never hope to restore. The utmost it can aspire to is to become the Chancellor of Nature's Exchequer, and to control for its own ends the immense reserves of energy which are at present in keeping for great cosmical schemes.

It looks, therefore, as if our successors would witness an interesting race, between the progress of science on the one hand and the depletion of natural resources upon the other. The natural rate of flow of energy from its primary atomic reservoirs to the sea of waste heat energy of uniform temperature, allows life to proceed at a certain pace, sternly regulated by the inexorable laws of supply and demand, which the biologists have recognised in their field as the struggle for existence. The flow of energy is, however, not a simple, straightforward affair, but proceeds in stages through intermediate reservoirs. The main part that concerns life on this planet is received as radiant energy. Part suffers useless conversion into its final form directly; a smaller part produces, through the evaporation of the ocean and rainfall, the "white fuel" or water power which is more and more being turned to practical use; yet another part in bygone days was stored up in the remains of the forests of the carboniferous era, and, in the form of coal, furnishes the main supply in use at the moment. Natural gas and oil may represent a residue of the initial heat energy of the earth, or even a portion of that continually being evolved by the radio-elements. At very high temperature, only to be artificially realised in the electric furnace, carbon enters into a series of compounds with the metals known as the carbides, which are decomposed by water, giving hydrocarbons. Acetylene is merely one instance. Moissan considers that the natural deposits of petroleum and supplies of natural gas have their origin probably in the decomposition, by the action of percolating water, of carbides formed deep down in the earth.

In so far as it has been found possible to accelerate the pace of life, beyond that fixed by the laws of supply and demand of energy, it has been by utilising these accumulated supplies of energy. As already remarked, the utilisation of water power is, in the energy balance-sheet, pure gain. But the small proportion of the increased demands of civilisation so furnished is negligible as compared with the enormous part from the consumption of coal and other natural fuel. It might be supposed that an almost unlimited expansion would be possible, as the demand arose, in the utilisation of the direct heat of the sun or of the energy of the tides. But it is doubtful whether either of these will ever suffice to drive directly a practicable engine. In their indirect application, in the improvement of agricultural processes by irrigation, in forestry and so on, no doubt the future will derive more benefit from them than the present. Sooner or later, but certainly not indefinitely later, nothing known will remain to supplement the natural rate of supply of energy, save the primary stores of atomic energy on the one hand and the waste heat energy of uniform temperature on the other—

a state of things that might aptly be represented by the proverb of the devil and the deep sea, so far as any existing knowledge goes. It is probable that the first of these alternatives is the less hopeless, and that the problem of artificial transmutation will in the future come to be regarded, no longer in the light it was a few years ago, as an impossible chimera surviving from the discreditable epoch in which the science of chemistry originated, but as the final phase of the agelong conflict of interests between Nature and Man. Success would remove for ever the physical limit to the continuous advance of progress, and would endow it with a permanent significance which at present it does not possess. Failure, on the other hand, would mean a gradual future relapse of the race into a more primitive condition, and the loss of much, if not most, of what distinguishes life to-day from that of our unscientific ancestors.

It has been stated that a problem defined is, in science, a problem more than half solved. The process of mental preparation necessary, before it is clear that a definite problem exists, often requires longer than the solution of the problem, once it has been recognised. Possibly the difficulties in the way of transmutation may prove less formidable than now appears. Time alone will answer that question. But it is characteristic of the age that a problem at first sight of purely philosophical character, whether matter is fundamentally one or of many different kinds, should have become, suddenly at the beginning of this twentieth century, the problem of the relation between Man and external Nature in its final and most fundamental form. No one to-day is ignorant of the part played by energy, not only in science, but in industry, politics, and the whole scheme of human welfare. From the cradle to the grave every one is dependent on Nature for an absolutely continuous supply of energy in one or other of its numerous forms. When the supplies are ample there is prosperity, expansion, and development. When they are not, there is want. Often, it is true, energy appears to play a very subsidiary and indirect part in the development, just as, no doubt, the supply of wind might be looked upon as playing a very secondary rôle in the music of an organ. The fact remains that, if the supply of energy failed, modern civilisation would come to an end as abruptly as does the music of an organ deprived of wind.

Before science had advanced to the knowledge of the utilisation of the stores of energy in fuel, the favourite method employed by earlier civilisations to augment their normal supplies of available energy was by means of slaves. To their skill in the utilisation of this form of energy the Pyramids of Egypt bear witness. To a negro the electric car appeared in the light of the emancipation of the mule. A single machine nowadays customarily does the work of 10,000 horses, good typical British cart horses, and accomplishes more work than a whole army of slaves could do. We are fortunate to be living in the days of cheap coal. It has but to be kindled when the sunshine of forgotten times does, at will, the work of the world, and the helot and the negro walk free. If we pause but for a moment to reflect what energy means for the present, we may gain some faint notion as to what the question of transmutation may mean in the future to a fuelless world, once more dependent upon a hand-to-mouth method of subsistence. It may still be centuries before this occurs, but neither the application of the discoveries of science nor even their achievement is to be compared with the struggle in winning them. It is a satisfaction peculiar to the present age to have learned that no physical poverty of Nature bars the road stretching away into the future. The world is great enough and rich enough to supply human aspirations and ambitions beyond all present dreams. But the human intellect must keep pace in its development with the expanding vision of natural abundance.

BIBLIOGRAPHY

Lectures on some Recent Advances in Physical Science, by P. G. Tait (Macmillan & Co.), though first published in 1876, gives a very valuable account of the development of the Doctrine of Energy during the middle of last century.

The Elements, by Sir W. A. Tilden (Harper & Bros., Library of Living Thought), 1910, deals with the Periodic Law as it appears to the trained chemist.

Brownian Movement and Molecular Reality, by Jean Perrin, translated by F. Soddy (Taylor & Francis, London), 1910, gives a vivid though, necessarily, not entirely non-mathematical account of the way in which the Brownian Movement has bridged the gap between Molecular and Thermodynamical Science.

Electrons, by Sir Oliver Lodge (Macmillan & Co.), 1906, and *The Electron Theory*, by E. E. Fournier d'Albe (Longmans, Green & Co.), 1906, may be recommended as introductions to the newer views in the subject of Electricity and Magnetism.

The Corpuscular Theory of Matter, by Sir J. J. Thomson (C. Scribner's Sons), 1907, gives an authoritative account of the author's experiments and attempts to found an Electrical Theory of Matter.

Light, Visible and Invisible, by Silvanus P. Thompson (Macmillan & Co.), 1897, contains a popular account of the nature of light and X-rays.

The Interpretation of Radium, by Frederick Soddy (G. P. Putnam's Sons), 1909, deals popularly with the phenomena of radioactivity, and their explanation by the Theory of Atomic Disintegration.

Inorganic Evolution as studied by Spectrum Analysis, by Sir Norman Lockyer (Macmillan & Co.), 1900, and *Worlds in the Making*, by Svante Arrhenius, translated by H. Borns (Harper & Bros.), 1908, discuss some of the problems of cosmical evolution, the first from the spectroscopic, and the second from a general point of view.

Radioactivity and Geology, by John Joly (D. Van Nostrand Co.), 1909, is a fascinating volume, showing the inter-connection between the two sciences indicated by the title.

INDEX

Made in United States
Troutdale, OR
11/03/2026

56845236R00146